THE SACRED MONKEYS OF BALI

THE SACRED MONKEYS OF BALI

BRUCE P. WHEATLEY
The University of Alabama at Birmingham

Long Grove, Illinois

For information about this book, contact:
Waveland Press, Inc.
4180 IL Route 83, Suite 101
Long Grove, IL 60047-9580
(847) 634-0081
info@waveland.com
www.waveland.com

Cover Illustration: An early engraving of *Macaca fascicularis* from Schreber, 1774, *Säugthiere*, Vol. 1. Note the subject's commensalism—holding a papaya in front of a house.

Frontispiece: An alpha female poses with a cooing kidnapped infant from another troop.

Chapter Opener Photographs: p. 6 Balinese temple; p. 38 the monkey dance, customary theater, Bali; p. 66 rice terraced landscape, Gung Kawi; p. 120 women placing offerings at temple, Ubud; p. 146 women at temple festival, Ubud.

10-digit ISBN 1-57766-059-5
13-digit ISBN 978-1-57766-059-0

Printed in the United States of America

9 8 7 6 5 4 3

To my wife, Cathleen,
and our three daughters,
Farrell, Lesley, and Bonnie

Contents

Acknowledgments

I am very grateful to the people and country of Indonesia for the opportunity to conduct research in Indonesia. I thank the Indonesian Institute of Sciences and the Research and Development Centre for Biology for sponsoring my research. I also thank Udayana University and the University of Indonesia. The following individuals deserve my thanks: D. K. Harya Putra, Moertini Atmowidjojo, T. Hainald, E.K.M. Masinambouw, Dedi Darnaedi, Jito Sugardjito, Soetikno Wirjoatmodjo, Jatna Supriatna, I. Gde Suyatna, Alit Widjaya, Soetoto, Arie Budiman, Soewarno, Wayan Atjin Tisna, Nyoman Sudhi, Nyoman Sugitha, I Made Sada Artha, Mrs. Artini, and Orville Smith. I am very grateful for the support of the Fulbright Program and the Center for Field Research. I am also thankful for the grant support of Earthwatch and its many tireless volunteers who made this research successful. Thanks also go to James Flege and the Department of Biocommunications at the University of Alabama at Birmingham for the loan of the spectrograph (NIH grant NS 20572). I also wish to express my gratitude to my two research assistants from the University of Alabama at Birmingham, Katy Gonder and Tripp Holman, and to my wife, Cathleen Wheatley, who collected data with me one summer in Bali and who also helped me write this book.

I wish to thank the Republic of Palau, especially David K. Idip, Ben Gulibert, and Riosang Salvador, for their help and permission to conduct research on Ngeaur. The assistance, help and friendship of Rebecca Stephenson and Hiro Kurashina were invaluable, both on Guam and on Ngeaur. The University of Guam kindly offered financial and other support, as did Continental Airlines. I also express my gratitude to the people of Ngeaur for their support. I thank the following students at the University of Guam who participated in the research: Marcellus Akapito, Robin Barrett, Jenny Chargualaf, Wilda Johnson, Kelly Kautz, Jennifer Matter, Yvonne Singeo, David Tibbbetts, and Cathy Ogo.

Acknowledgments

[illegible]

[illegible]

Introduction

Anthropology is a rich and diverse area of research that utilizes many different approaches to the understanding of humanity. Each of the subfields in anthropology focuses on a particular aspect of humans, and the discipline's emphasis on fieldwork adds another dimension to the richness of the field. As a primatologist who studies monkeys, my subfield is physical anthropology, the study of human biology. Although my training was in all of the four subfields of anthropology—cultural, linguistics, archaeology, and physical—I began my career studying monkeys as animals, albeit very interesting animals.

Watching monkeys or more accurately, macaques, in Kalimantan or Indonesian Borneo for two years, I began to realize the importance of human influences on these animals, such as the impact of slash-and-burn or swidden agriculture. Studying the same species of macaque, *Macaca fascicularis*, once again in Java and Bali, I began to realize the depth and the importance of the culture's heritage about monkeys and how different it was from my own.

In the last twenty years of teaching cultural and physical anthropology, I found myself straying more and more "over the line" into cultural anthropology. The area where humans and monkeys intersected was simply irresistible, and I did not see too many cultural anthropologists doing it. I still feel somewhat guilty about this transgression, and I have to say that I am not a cultural anthropologist. A trained cultural anthropologist specializing in that field would, no doubt, do a much better job than I regarding the Balinese culture would. Cultural anthropologists may, in fact, find some ideas and interpretations in this book that are objectionable or simplistic. I can only hope that they will be tolerant and practice a little "subfield relativism" on my attempt to show the importance of Balinese cultural attitudes toward monkeys as they relate to the development of conservation at one of the monkey forests in Bali.

These are difficult times for our time-honored four-field approach. The American Anthropological Association, for example, recently calculated that only 28 percent of all university anthropology departments have faculty in all four subfields (Morell, 1993). Departments are dividing where the schism is especially wide: between the physical or biological subfield and the cultural subfield. A recent survey of physical anthropologists revealed that "without doubt" the traditional four-field approach was no longer appropriate (Wienker

& Bennett, 1992). Cultural anthropology, for example, was thought to be of only marginal importance to physical anthropologists. Annette Weiner (1995) and James Peacock (1997) both addressed this problem of the split-up of our field in their recent presidential addresses to the American Anthropological Association on the future of anthropology. They emphasized the need to merge the insights of our fieldwork with our theory in order to solve society's problems and to be relevant. According to Peacock, if we are to be successful in bringing our discipline out of its present "marginal category," then we need to "use it or lose it." Our separate but equal subfields should learn from each other and come together. We could then integrate our understanding of the world and begin to shape it. Peacock (1997) called for a multidisciplinary approach to enhance human welfare.

This book was written with this multidisciplinary and mutualistic spirit in mind. I think that the separation and antagonism between the subfields is a mistake. One of the rationales of primatology, the study of primates, is that we can understand ourselves better. Primates have been used as models for humans with remarkable success in the area of biomedicine for a long time. The development of many vaccines and the discovery and understanding of the Rh (initials deriving from rhesus monkeys) factor in humans are just a few notable factors. Sherwood Washburn and Irven DeVore (1961) also utilized this approach in their famous studies of baboons on the African savanna. Such studies help us in our hypotheses about human evolution by making us consider how natural selection might operate on the behavior of baboons and, by inference, early humans in an environment where we evolved.

My approach is broader. I am interested in pursuing questions and answers to every area in which nonhuman primates intersect with us. For example, what can primatologists offer to cultural anthropologists? How might a study of monkeys help us understand Balinese culture? Could an understanding of Balinese attitudes toward monkeys help us understand ourselves, as Westerners, or ecologists? One of the most important areas in which anthropology can be relevant is the field of conservation. The collaboration of physical and cultural anthropology to solve the problem of habitat loss and endangered species typifies what each of our subfields in anthropology can do for our welfare. Thus, while this book is about a species of macaque, it is also about my collaboration with local scientists and local people for a common goal, namely, the conservation of the Monkey Forest. The application of this collaboration toward conservation goals may prove to be a useful model elsewhere in the world. This approach is also a good example of what the American Anthropological Association recognizes as our fifth subfield, applied anthropology.

I first went to Bali to watch monkeys in 1976, but it was not until 1986, after my Fulbright Award to Indonesia, that I had a chance to return to Bali. My family and I spent a year living in Jakarta, the capital of Indonesia and one of the world's largest cities, while I taught at the University of Indonesia. At the end of the Fulbright my family went home with a new addition, Bonnie, who was born in Jakarta in 1986, while I went to begin my research in Bali. With the first of my three Earthwatch Research Grants on Balinese Temple Monkeys in 1990, I tried to persuade my family to return to Indonesia once again. Bali was different from the urban center of Jakarta, I told them. The scenery was beautiful, with volcanoes, lakes, temples, beaches, cliffs, people, and monkeys. Even the weather in the summer was better in Bali than in Birmingham, Alabama. My three girls asked why we couldn't be like normal families and spend the summer in Disney World, or even Hawaii? Who among their friends had ever gone to Bali or had even heard of such a place? Finally, after more slide shows, my family went to Bali in 1991. Needless to say, we had a wonderful time, and none of us needs any more persuasion to return to Bali.

What do the Balinese think about their monkeys? How important are monkeys to the Balinese? I suppose that if we conducted a survey, we would find that some of the people like monkeys and that other people don't, just as we might if we surveyed Americans. Imagine my surprise when I realized the possibility that Bali was named after a monkey king. To my knowledge, no other anthropologist has ever discussed this possibility! I therefore begin this book with this idea and with this illustration of how the subfields can mutually profit from each other.

The book begins by examining the important role that monkeys play in Balinese culture. Chapter 1 discusses the impact of the Ramayana Kakawin, the ancient and holy Javanese version of the Indian epic poem, on the religion and arts of Bali. It is in this and other poems and stories that the derivation of the name *Bali* may be found. This chapter also differentiates Balinese views of human nature from Western views and discusses how these views affect attitudes toward monkeys. Westerners tend to think of polar opposites with a sharp dichotomy between human reason and morality on the one hand and the sinful and evil monkey on the other hand. Balinese can accommodate such polar opposites, however, and are able to integrate them. Thus, the Balinese can have, as I argue, not only a godlike monkey, such as Hanuman, but also an animal-like and demonic monkey.

Chapter 2 discusses the antiquity and prevalence of human-monkey interactions or primate commensalism. Humans have modified every ecosystem on earth, and we can no longer treat such influences as well as the animals in them, as "unnatural." Chapter 3 presents new and detailed data gathered within a seven-year period on the behavior of the long-tailed macaque in the Monkey Forest of Padangtegal near the village of Ubud, Bali. For the first time in this species, the behavior of females as well as intertroop behaviors are elucidated. Chapter 4 discusses how Ubud, one of the primary targets of the ecotourist, is one of the few communities in the world that has managed to control development and use tourism to benefit the local community. This chapter shows how the local people developed a conservation plan and how I worked with them to manage the Monkey Forest. The final chapter discusses the integration of subfields into a new symbiotic field that I have called *cultural primatology*. This field promises to be a rich area in the investigation of the interactions between human and nonhuman primates.

Chapter 1

Bali, the Monkey King

The Island of Bali

Bali is a small island, ninety miles long and fifty miles wide, located 600 miles south of the equator, in the middle of the world's largest archipelago. The island is shaped in the form of a rooster. Some Balinese say that this shape is appropriate because of the important uses this animal has in sacrificial offerings and cockfights. It is only one of about 17,000 islands spanning some 3,200 miles east to west that make up the country of Indonesia. Bali became one of the sixteen provinces of Indonesia in 1958. Its estimated 1996 population was 2,924,400, with a density of 526 people per square kilometer (*Europa World Year Book*, 1998). It contains many if not most of Indonesia's Hindus with 93.2 percent of its population. In 1991 the percentage of Muslims was 5.22 percent, Protestants .58 percent, Buddhists .55 percent, and Roman Catholics .47 percent (Frederick & Worden, 1993).

Although Bali is a small island, it makes a lasting impression on visitors. The most conspicuous physical features on Bali, as almost everywhere else in Indonesia, are the volcanoes. Bali has an arc of volcanoes in its central and eastern section, approximately six of them over 2,000 meters tall. At 3,142 meters, or 10,308 feet, the highest volcano is Mount Agung, meaning great, or majestic, mountain. Nearby Mount Batur is 1,717 meters. Even today, steam is visible from its peak. These volcanoes are part of the Batur caldera. The god Siwa is said to have pushed apart the cosmic mountain, creating Mount Batur and Mount Agung, thus giving rise to the split doorway so commonly visible in front of every temple. These and other volcanoes have great cosmological significance because they are a source of destruction and fertility. Eruptions, earthquakes, and mudslides are frequent and deadly.

For example, one of the worst earthquakes took place in 1917, on January 21, destroying 2,500 temples and 65,000 homes. Ubud was completely flattened (Sukawati, 1979). More people, however, actually died from the epidemic that followed than from the eruption of Mount Batur and the earthquake. Balinese volcanoes also spew ash, pumice, and water, making the soil fertile for wet rice agriculture, or *sawah*. E. Mohr (1945), for example, explicitly linked higher population densities with the presence of active volcanoes in Indonesia. Heavy rains also leach the nutrients from the soil, especially during the tropical monsoon season from December to February, when on average there are twelve inches of rain per month. Such a situation almost seems to require a good volcanic eruption from time to time. The dry season lasts from April to November with an average of only four inches of rain per month. The most important eruption in recent times was that of Mount Agung in 1963. At that time, rain fell that burnt the skin as well as the foliage (Mathews, 1994). Over 1,500 people died, with almost 100,000 left homeless. Just as in some previous eruptions, for example, the one in 1815, when Mount Tambora on Sumbawa erupted and destroyed crops in Bali, the 1963 eruption was also tied to political upheavals leading to great loss of life in 1965–66. Balinese volcanoes are quiet and beautiful for long periods of time between violent eruptions.

The country of Indonesia may at times have been less known to Westerners than the Island of Bali. Indonesia is the world's largest Islamic country, with 90 percent of its people Muslims and most of the rest Christians, with a small percentage of Hindus, Buddhists, and animists. It is the fourth most populated country in the world, with 209,565,000 inhabitants as of 1997 (Banks & Muller, 1998). Indonesia is home to more than 250 ethnic groups and cultures and is thus one of the world's most diverse cultures. Indonesia is the most heavily forested Asian country and the world's largest exporter of plywood and rattan. It is also the world's fourteenth largest oil producer. Some of the largest and most diverse flora and fauna in the world, including many of the world's primates, are located in this country. Indonesia spans several major biogeographical regions where, for example, Wallace's Line runs through the middle of the country (see map), dividing the islands of Borneo and Bali from Sulawesi and Lombok (Walker, 1980). Indonesia is also one of the homes of one of the world's most numerous and widely dispersed nonhuman primates, *Macaca fascicularis*.

One of the most important influences on Bali and Indonesia was India. The name *Indonesia* itself reflects its Indian heritage, because it is a shorter synonym for the Indian Islands or the Indian Archipelago. Indonesia thus means the Islands of India, and its people and cultures were thought by Europeans to be offshoots of India. The area has been called the Indies at least since the sixteenth century, when spices were highly desired in Europe. George Windsor Earl, in an article with James Logan, coined the word *Indonesia* in 1850 in the *Singapore Journal of the Indian Archipelago and Eastern Asia* (Jones, 1974).

The word *Bali* also reflects its Indian heritage. *Bali* is a very ancient Sanskrit name, much older than the name *Indonesia*. Beyond this one the Island of Bali has many names. To many Indonesians, Bali is known as *Pulau Dewata*, or The Isle of Light, from the Sanskrit *dwy*, meaning light. This Sanskrit heritage from India was invoked by the late prime minister of India, Jawaharal Nehru, when he called Bali "the last paradise, the morning of the world" when he visited the island in June of 1950 (Last, 1994). This metaphor is particularly meaningful among the many metaphors applied to Bali. When dawn strikes most of the population of Indonesia, it is already morning in Bali. In Bali, the sun rises over the holy slopes of Mount Agung. Just as the the sun rises in the east, Bali is forever fixed as a rare and shining jewel to the east of Java. However, other Balinese principles rather than just this horizontal orientation, are also invoked, such as a vertical orientation with the gods, various colors, in this case white, and the association of the sunrise with birth and the beginning of the day for humans. Perhaps too, as Nehru said, Bali is the hope or paradise of the world.

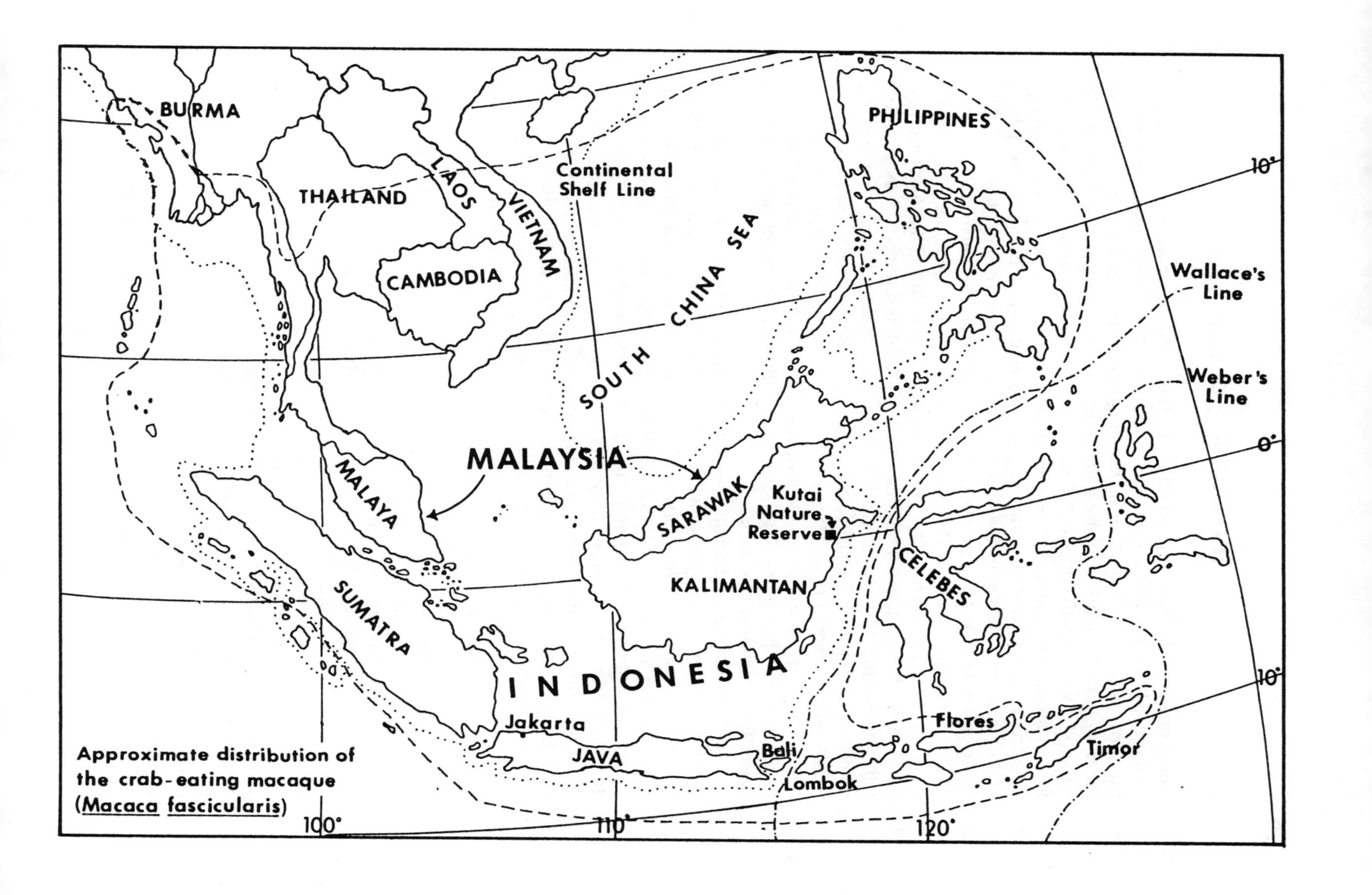

Approximate distribution of the crab-eating macaque (Macaca fascicularis)

The Naming of Bali

The word *Bali* in Sanskrit means offering, tribute, propitiatory oblation (Atmodjo, 1974; Suchitra, 1994; Zoetmulder, 1982). Another name for Bali is *Banten*, which is the Krama, or higher-language-level term for Bali; and it also means offering, usually to the gods above. How and why did Bali get its name(s)? What was the offering? There are several explanations. Soewito Santoso (1980a), the translator of the Ramayana Kakawin, which is the Javanese version of the Indian Ramayana story, as well as other epics, says, "the Island of Bali is in fact called after the brother of Sugriwa, the King of Monkeys." The name of Sugriwa's brother mentioned in the Ramayana Kakawin is Bali, a powerful and evil monkey king. Bali becomes evil only when he takes his brother's consort, Tara. The hero, Rama, saves the world by killing Bali, to obtain Sugriwa's help. Bali is thus "the buffalo to be slaughtered and used as an *offering*"(italics mine) in the battlefield for Sugriwa (Santoso, 1980a). Rama is the officiating priest, and Tara, the wife of Sugriwa, is "the fortune of the offering which Sugriwa would relish" (Santoso, 1980a). When Bali is dying, he says he hopes to be a good brother in the future and live in the mountains with lots of ripe fruits. He then gives Sugriwa a golden flower, a symbol of his priestly power from his head, and dies.

Bali, usually called Subali, is therefore a wicked monkey in this story, but there is another, more sympathetic story that more fully explains his unfortunate situation (de Zoete & Spies, 1938; Dibia, 1979). In the Kuntir Legong story, the god Indra goes to Mount Semi and asks two monkey priests named Subali and Sugriwa to help him fight a demon. He promises his daughter, Devi Tara, to the one who succeeds in getting rid of the demon. The demon, Mahesa Sora, runs into a cave on Mount Himawa, and before Subali enters to fight him, he tells Sugriwa to guard the entrance. Sugriwa will know the outcome of the fight by watching the color of the blood flowing out of the cave: red for the demon and white for Subali. Sugriwa sees both red and white flowing out of the cave and, thinking Subali dead, seals up the cave with a boulder. Sugriwa receives Indra's daughter, Tara, as his wife.

Subali, however, is not dead. The white color was from the demon's brains, and when Subali finally escapes from the cave, Indra tells him that he has an equal right to Tara. Subali defeats Sugriwa, takes Tara, and they live happily until Sugriwa tells Hanuman to find Rama, who also needs help in obtaining his wife, Sita. Rama kills Subali, of course; and Sugriwa, together with Hanuman, helps Rama find Sita. Subali thus appears in every sense a nobler monkey than Sugriwa (de Zoete and Spies, 1938). His predicament is the result of a misunderstanding. By becoming a sacrificial offering, Bali is an essential component in the restoration of harmony in the world. The fact that these two monkeys are priestly yogins indicates their harmonious relations to the deities. The gifts obtained through their meditation are force, extraordinary powers, many relatives, being a refuge for people, the ability to see as near what is far off, and knowing the heart of other people (Hooykaas, 1964). The white blood of Bali is reminiscent of that of King Judistira, the eldest of the Pandawas, who was so truthful that the gods made him the only human whose feet never touched the ground (Moertono, 1981). Forests, mountains, and caves are powerful places, where yogins obtain inner strength. The flower from Bali's head may represent his life force, and it may be similar to that of priests: a Buddhist priest wears a flower over the ear, whereas a Siwiast priest may have a flower in his hairknot. Is this exorcism? Is Bali a symbol of the fusion of godly and natural worlds? Is this story an anagram or antidote for saving the human world?

R. Friederich (1959 [1849–50]) gives another explanation for the naming of Bali. From the ancient Bali Sangraha, a manuscript that no longer exists, *Bali* means *wiseha*, and *sangraha* means *kumpulan*, thus "the gathering of the excellent (the heroes)." Friederich says, "Bali is then not to be considered as 'offering,' but as the nominative of *balin*, a strong person, powerful, a hero." The word *balin*, however, is another name for Bali, the monkey king (Dowson, 1972). Dowson goes on to say that Bali's father was Indra and that the name Bali derives from *bala*, or hair of his mother from whom he was born.

These stories and many others are part of the epic poem, the Ramayana, known by most of the population in Indonesia, especially on the islands of Java and Bali, as well as by most South and Southeast Asians (Scott-Kemball, 1959; Zoetmulder, 1974). One can see reliefs of the battle between the two monkey-king brothers (Sugriwa and Balin) on temple walls in Java, for example, the Candi Loro Jongrang, and in Cambodia on the Banteay Srei near Angkor Wat (Zimmer, 1968). Statues of Rama and Hanuman, the monkey general, both central figures in the Ramayana story, can be found at a twelfth-century Balinese temple complex called Pura Sada (Atmodjo, 1974; Grader, 1990). The rivalry between Subali and Sugriwa is told in the dances of Bali, such as wayang wong, legong, jauk, and baris (de Zoete and Spies, 1938). Drawings of Bali appear to be rare. Hedi Hinzler (1983) describes one such drawing of a monkey named Bali sitting next to Rawana plunging into the sea. This drawing is by the dalang (puppeteer) Ida Made Tlage, who bases his drawings on traditional sources. Whatever the real truth behind the naming of the island of Bali, it is essential to take a good look at the arts and the Ramayana.

The Arts of Bali

The Indian influence on Bali is much more than the name of the island. Evidence of trade between India and Bali and between India and Java goes back to at least the first two centuries A.D. (Ardika & Bellwood, 1991; Walker & Santoso, 1977). Distinctive Indian pottery with Kharoshthi script has been found in northeastern Bali and in Java. Contacts between Bali and Java were extensive. The Old Javanese language, with half of its words in Sanskrit, had replaced the Old Balinese language by the tenth century in Bali (Goris, 1960). Intermarriage between the dynastic families of Bali and Java occurred, and many Old Javanese literary manuscripts, possibly even the Ramayana Kakawin, also entered Bali by this time. Javanese influence in Bali increased further when Bali was defeated in 1343 and formally became a political part of the Majapahit Empire. This date separates the earlier Hindu-Balinese period from the ensuing Hindu-Javanese period, according to P. J. Zoetmulder (1974).

By whatever route it came, the Indian influence can be seen in many areas of Balinese life, from Hinduism, its caste system, its bureaucratic administration and Code of Manu, writing, manuscript preparation, and the introduction of numerous Sanskrit words (Sankaranarayanan, 1992). These aspects of Indian influence on Bali are just that, influences. Balinese Hinduism, for example, is very different from Indian Hinduism. But above all, it is in the ubiquitous and spectacular arts that this Indian influence is so prominent, such as in dance, the palace and temple architecture, stone and wood carving, court theater such as wayang or shadow puppetry, drawing, painting, and batik (Rassers, 1925). Every village in Bali strives to have its own dance troupe, gamelan orchestra, painters, sculptors, and so on.

It is sometimes difficult for Westerners to understand why the Balinese are so consumed by the arts. It is true that many of its arts, such as the dances, are modern and performed for tourists, but we would be very wrong to think that their main purpose is to make money. This commercialization does not make their culture a touristic one. The Balinese enjoy these performances too. The number of wayang or puppet performances in Bali is greater today than ever before, despite their lack of popularity with tourists (Zurbuchen, 1987). Many of the themes expressed in the modern dances, shadow plays, theater, manuscripts, paintings, and sculptures of the villages derive from the classics, such as the epic Indian poems, the Ramayana or the Mahabharata. It is the continual variation of the themes from these epics that makes Balinese culture so vigorous with respect to the small size of the population and island (Marrison, 1986). Change will not destroy Bali. It is the absence of change, the fossilization of Balinese culture, that is the greater danger (Pollmann, 1990).

Bali is a theater state (Geertz, 1980) because the arts of Bali sustain its Hindu-Buddhist civilization not only by forming the means of its transmission, but also by maintaining order through the activation of temple ceremonies (Lansing, 1983). It is through the arts that the appropriate relationships between gods and humans are maintained. A harmonious balance must be struck between these various worlds to keep order in the world we live in. The fact that tourists might pay for and enjoy these performances is an added bonus.

Before I discuss the Ramayana Kakawin (the Javanese epic poem of the Ramayana story) in more detail, it is necessary to define and describe what a kakawin is and why it is holy. This will also help explain how the arts are used to maintain order. A *kakawin* is a poem, a "language monument" written by a *kawi* or poet (Zoetmulder, 1974). It is the abode and monument of a king or deity that mediates between the two worlds (Henry, 1987). The poets who "built" poems and other forms of literature are "priests of the magic of literature" (Zoetmulder, 1974). The effect of a poem is a sense of oneness, a loss of consciousness or trance. The "two-sided" word *lango* is used to describe this state of rapture, in which the subject is completely absorbed by and becomes lost in its object (Zoetmulder, 1974). *Lango* means enraptured and enrapturing. The word has both a subjective and an objective aspect. A kakawin is thus a result of the poet-priest's unification with god, and it is a receptacle of the god, not just for the poet's meditation but also for the poet's audience.

It is in this way that a kakawin becomes the poet's temple, in which he unites with god and achieves ultimate liberation and release from reincarnation. The script and alphabet used in the kakawins are sacred in Bali. When the Ramayana Kakawin is chanted, the Balinese ancestors are believed to be talking. Poets of Kawi (Old Javanese) are revered as representatives of the supreme deity as progenitor and patron of the arts (Hunter, 1988). The poet brings the creative work of the deity alive by uniting word, image, and language. Balinese dalangs (puppeteers) open their performances by stating that the performances are based on these holy Sanskrit texts, and the correct order of their recitation helps keep stability in the cosmos (Hinzler, 1993). The holy Ramayana Kakawin manuscripts are kept in special carved and painted lidded wooden boxes (Hinzler, 1993).

Another important kakawin, in addition to the Ramayana, is the Sutasoma. The motto of Indonesia, "Bhinneka Tunggal Ika," or Unity in Diversity, is taken from this fourteenth-century Old Javanese kakawin. This kakawin is important in Bali for its synthesis of Siwa and Buddha.

Priests meditate at certain places. One such place near Ubud, Bali, is a beautiful spot called Mount Kawi, the poet's mountain. Some of the temples at Mount Kawi were built in 1080 as a monument to Anak Wungsu, one of Bali's greatest kings. The ashes of his royal remains are said to be here, and the temple complex, including a hermitage for priests, is a site for the health and well-being of pilgrims and for the community (Goris, 1960; Kempers, 1991). This area is a center of mediation between the gods and humanity. There are many such areas all over Bali, and the temples frequently have Ramayana scenes carved in their walls. Pura Puseh, a temple in Kapal, for example, has scenes of the Ramayana, including those of monkeys (Hinzler, 1990). Fifteen kilometers away is Sangeh, the site of an ancient and sacred monkey forest. According to the Ramayana, the monkeys in this nutmeg forest were flown here by Hanuman, the Monkey Chief, along with the cosmic mountain (Hinzler, 1990).

The Ramayana

The Ramayana story is one of the oldest written accounts of human and monkey mutualism. It is the most popular and beautiful piece of literature of ancient Indonesia (Santoso, 1980a; 1980b). It is also sacred and meant to be chanted or sung (Uhlenbeck, 1989). About 30 percent of the words in this poem are of Sanskrit origin. The script also originated from South India. The Indian version of the Ramayana was written around the first century B.C., but it was composed much earlier, perhaps in the fourth century B.C. (Marrison, 1986). The oldest surviving Indonesian version of the Ramayana, the Ramayana Kakawin, is commonly believed to have been written in King Balitun's reign

(A.D. 898–930) of the Mataram Kingdom in Central and East Java, but it may be even older, perhaps before 760 (Santoso, 1980b).

The Ramayana story pervades all aspects of Balinese life. Its reading is an important part of life-cycle rites such as six months after conception, tooth fillings, weddings, and cremations. It is a popular and well-known subject of paintings; it is associated with the courts, requiring an elaborate and expensive presentation, and its story is told in countless dances (Marrison, 1986). Monkeys are common characters in these stories. One such dance in particular, the *kecak*, is worth mentioning in greater detail.

The kecak is a very popular drama sometimes called the monkey dance or chant (see Plate 1.1). It sometimes has 100 male vocalists singing with interconnecting syllables. They chant CAK, CAK, or CHACK. This component of the kecak goes back to the more mystical and traditional dance called the *sanghyang*, in which a male choir formed the vocal background or, cak

Plate 1.1. *Hanuman dances at a local performance of the kecak in Padangtegal.*

(Moerdowo, 1958). In many villages in 1948 the sanghyang was danced for months after guerrilla warfare in order to restore equilibrium between good and evil. One version of the kecak that many tourists attend was developed in 1930. It blended a short Ramayana dance/story with a voice rather than instrumental gamelan. Some villages included the story of Subali and Sugriwa in their kecak (deZoete and Spies, 1938).

The kecak became popular in the area, and the German painter Walter Spies commissioned performances for tourists in 1932 (McKean, 1979; Picard, 1990a). The kecak is a dance of exorcism where, for example, the dancer becomes the demon Kumbhakarna and the chorus becomes monkeys (de Zoete and Spies, 1938). President Achmad Sukarno, who had a Balinese mother, saw this performance in Bona in 1950 and asked that it be taught in the neighboring villages, believing that it would be a mechanism to unite villages, bring in tourists, and exorcize evil during times of peril, famine, and pestilence. Young girls would go into trances, dance on the shoulders of men, and request advice from the ancestors on how to cope with problems. The dance would go on for days, if not months. Colin McPhee (1948) and B.J.A. Lovric (1988) also describe sanghyang in which men, in trance, become monkeys or other animals. The dancer is transformed into the spirit of a monkey, and he bites and kicks like a monkey. The spirit is asked to go home at the end of the dance. Lovric (1988) goes on to make the connection between the movements in these dances and disease categories such as *tiwang*, or convulsive seizures. In other words, these sanghyang dances are pantomimic. They conjure the various spirits responsible for diseases and then dismiss them. Such exorcist dances enact their magic where rapport is established with the demonic in order to protect the community. It is not a question of good annihilating evil, says Lovric, but that of a restoration of balance of power for health.

The Ramayana Kakawin is also much revered in Bali because of its lessons in morality (Zurbuchen, 1987). The most important reason for this is that the story was written by a Brahman yogin (Santoso, 1980b). The aim of the Ramayana is the ultimate goal of life: *moksa*, or heavenly bliss. The simple act of reading this magical and powerful story gives one a pure mind and leads to heavenly bliss. This is called the fruits of hearing it, or *srawanaphala* (Zoetmulder, 1974). The characters themselves are also pure. Sita is the faithful wife; Hanuman is the wise, brave envoy devoted to Rama; Sugriwa wants to serve at Rama's feet; and Rama is also as pure as the depth of the sea. Rama is the protector of the earth, the patron of humankind; he eliminates the impurities of the world. Rama is the god Wisnu, the preserver who created the world and makes sure that good people prosper and evil ones perish. Evil demons are committed to the destruction of earth and humans, and Rama is committed to the destruction of evil. The Ramayana is therefore a text that can restore or produce purity.

The Ramayana is used for many different purposes today. It explicitly expounds on the duties of the king, the so-called Asta-brata, or eight virtues. These virtues are the product of the eight deities within the body of a king. They are also known as the Serat Rama that appear in many other Javanese manuscripts (Moertono, 1981). The virtues are unlimited beneficence, the ability to repress all evil, kind and just conduct, lovingness, keen awareness and deep insight, generosity, ruthless intelligence, and courage in opposition to enemies. These qualities of an ideal king give him the primary responsibility for the prosperity of the kingdom whereby all his subjects follow his lead.

That these qualities of the ideal king are important in Bali can be seen, for example, in the Babad Buleleng (Worsley, 1972) where the King, Panji Sakti, is described as meeting these criteria, accounting for the harmony and stability of his realm in north Bali.

The Ramayana also gives rationale for political stability and mobilization. Just as in India, Rama/Wisnu is seen as the protector of the realm and of Hinduism against foreigners (Pollock, 1993). The first Balinese ruler, Sri Kresna Kapakisan, is likened to the god Wisnu (Wiener, 1995). Brave warriors who die in battle are said to go to Wisnu's realm. Wisnu provides for stability, and his consort, Sri, the rice goddess, provides fertility to the realm. The unified Balinese Kingdom of Gelgel in the sixteenth century became the symbol of Balinese social, political, and cosmic order. The leaders of all later kingdoms validated their claims through ancestral lines stemming from the rulers of Gelgel (Creese, 1991).

The Gelgel King Baturenggong established a close relationship with Nirartha, the Brahman of Brahmins that together formed the ideal kingship. Baturenggong was a "fierce opponent of Islam" (Hagerdal, 1995). The King fought off foreign invasion and extended his power from East Java to Lombok and Sumbawa. Adrian Vickers (1989) likens the king as the world ruler, an incarnation or son of a Hindu god whose wife is compared to Sita, and they both lived in a palace at Gelgel that rivaled King Rama's palace in the Ramayana. There is also a connection to Hanuman in the form of Agung Maruti. It is a bit unclear whether Agung Maruti was Gusti Agung of 1665, the ruler of Gelgel, but he was a usurper. In any event, Wiener (1995) relates that the source of Maruti's power was his tail and that he resembled an animal. Maruti's defeat involved subterfuge from a beautiful woman who tells him that the "gods would not allow someone resembling an animal to sleep with a woman of her beauty" (Wiener, 1995). The name Maruti is another common name for Hanuman (Hooykaas, 1970; Zoetmulder, 1982) and is mentioned frequently in the Ramayana Kakawin.

A possible example of the Ramayana being used in Lombok just prior to the Dutch invasion and defeat of the Balinese rulers is mentioned in G. E. Marrison (1986). This is a manuscript copied in 1893 in the Lombok collection, which is described as being used by Anggada, who fought alongside Rama, to urge on the monkey army against the demon Kumbhakarna. Anggada is a monkey commander and son of Subali (Santoso, 1980a). The Dutch victory in Lombok was climaxed by a *puputan* (ritual suicide) of the princes of Lombok. Such events were repeated fifteen years later by the Dutch in their massacres in Bali against Badung and Klungkung (Vickers, 1989).

Modern rulers also appeal to the Ramayana. Sukarno, the Indonesian president, referred to it in a speech in Denpasar, Bali, in an effort to stop the revenge killings after the revolution in 1950. He reportedly said, "You want to be the heroes of the revolution, but where in your Ramayana or Mahabharata do you find heroes who murder in the dark with their daggers, and when did Krishna or Ardjoeno take revenge on children and women?" (Last, 1994). Another recent example of the integration of the Ramayana into Balinese nationalism is the ancient story of Jaya Prana (Franken, 1984; C. Hooykaas, 1958). The name itself, Jaya Prana, can perhaps be translated as "victorious inspiration." Jaya Prana loyally gives up his life to his foster father, a king, when the latter covets Jaya Prana's wife. She is likened to Sita, Rama's faithful wife, when she kills herself. Nature itself rebels against these deaths and the king eventually kills himself (Franken, 1960).

There is a reference in this story to the Monkey King Sugriwa and his brother Subali (called Bali in one manuscript) and also a reference to Rama (Hooykaas, 1958). This story swept up Bali in 1949 when a village in which Jaya Prana died decided to give him a cremation to quiet his spirit and to help effect a religious revival in the village. The villagers had suffered tremendously through sickness and poverty. A number of miracles were witnessed and Jaya Prana was equated to Nirartha, the Brahman priest who brought Hinduism from Java. Although Jaya Prana was a commoner, he was cremated as a Brahmin, and the ensuing celebration became symbolic of the dead freedom fighters against the Dutch. A poem written by I Ketut Puthra was published in 1954 comparing the noble hero to Rama, sent to Bali to fight evil.

Bali, itself, then arises directly out of the sacred texts of the Ramayana, in name, tradition, and morality. Despite its antiquity, it is a model for the living: its message is that humans are an integral part of the world and its order and that the highest good is to live in harmony with it (Nayak, 1986). The values and guide to social conduct presented in these sacred texts and traditional arts give us insight on how the Balinese accommodate new ideas from Western contacts. As one of the oldest written records of human and monkey mutualism, it is therefore appropriate that the message of the Ramayana is utilized to develop conservation at the Monkey Forest in Padangtegal, Ubud.

Worldviews

It is difficult for many Westerners to read the Ramayana without dismissing it as a misguided fairy tale, with its many depictions of gods and demons that transform themselves into fantastic and mystical creatures with equally fantastic and mystical weapons and powers. Westerners tend to view monkeys in very negative terms, as in Elton John's putdown of Keith Richards, from *Rolling Stone* (Nov. 27, 1997): "He's so pathetic—poor thing. It's like a monkey with arthritis, trying to go onstage and look young." Contrast this with a mantra used by *Balian Usada* (literate healers): "May the . . . monkey-god Sang Hanuman [dwell] in the right hand, the noble monkey Sang Sugriwa in the chest" (Hobart, Ramseyer, & Leeman, 1996). The Balinese, however, take the lessons from the Ramayana quite seriously. Can a monkey, such as Hanuman, be a god or godlike? What can we learn about ourselves from our knowledge of such Balinese attitudes? Before I answer these questions and return again to the Ramayana, I need to explain the different ways in which a culture perceives and relates to the world.

The cultural differences expressed above regarding monkeys reflect a difference in what anthropologists call worldviews. A culture's worldview is connected to its classification of reality and how it divides up the world into different areas, such as the human or natural environment. One of the goals of anthropology is to uncover these different beliefs of people. Anthropologists try to be objective by practicing cultural relativism; that is, we attempt to explain the practices of another culture within that culture's own system or framework. This attitude avoids ethnocentrism or viewing another culture—often unfavorably—from the viewpoint of one's own culture. This can be difficult when your own culture has only one predominant way of looking at something. Another goal of anthropology is to look at our own beliefs from the perspective of another culture and to see how our beliefs affect other aspects of our culture. For example, do the negative attitudes that we tend to have about nonhuman primates affect our thinking about their capabilities or intelligence? I think so. Let us examine how different ideologies affect our place in the universe. For simplicity, I will contrast a Western worldview with an Eastern worldview. A good place to look for the worldview of a culture is in its religion.

A Western View of the Universe: Humanity Apart from Nature

The historian Lynn White, Jr. (1967), views Christianity "in absolute contrast to ancient paganism and Asia's religions." The Christian worldview is derived from the biblical view in which God is hierarchically above humans, who are, in turn, above animals. The human sphere is sometimes defined as culture, whereas the animal sphere is defined as nature. The boundaries of Christianity cannot be crossed. Christianity tends to "drive the opposites apart or to annihilate one side" (Ohnuki-Tierney, 1987). Eastern religions, however, tend to emphasize the synthesis of two opposing principles, such as the yin and yang of China and Japan (Ohnuki-Tierney, 1987) that bring opposites together in a "give and take" without being absorbed (Sharon, 1993). In many Asian philosophies, including Indian and Balinese, sharp boundaries between dichotomies do not exist because humans consist of the same five elements that make up everything in nature and the cosmos. Eastern doctrines allow transgressions of opposites but require mediation.

These viewpoints have a great bearing on how people relate to the world, especially in the realm of conservation. For White (1967), Christianity "insisted that it is God's will that man exploit nature for its proper ends," and he argues that it is no accident that the technological revolution was driven by Christianity, under which humans were now the exploiters of nature rather than just a part of nature. Ole Bruun and Arne Kalland (1992) state that the domination model of Judaeo-Christian cosmology where humanity is master over nature is responsible for many of our environmental and ecological problems. The relationship between humans and nature is the opposite in many Eastern cultures, where there is a "harmonious unity of mutual respect." They have more empathy with nature. The incorporation of humanity "back" into the natural world is an important step toward realizing conservation efforts. Contrast the Asian empathy of man and nature to the "modern" or Western conservation efforts that, in order to save nature, we need to take humanity out of nature. For example, Anita Pleumarom (1994) relates a story about how a group of people was forced off their land in the name of conservation and told not to hunt animals. One man was said to have remarked, "They say they want to protect nature. But aren't people also part of God's nature?" In Thailand, the World Bank is providing millions to conserve, manage, and develop forests and wildlife sanctuaries. Told to leave his ancestral home, a Karen tribesman said that if their ancestors had not done such a good job protecting their forests for centuries, then they wouldn't be faced with eviction (Pleumarom, 1994). Traditional Balinese architecture integrates man with the natural world: buildings seem to fit into their natural surroundings (Budihardjo, 1986). Space, land, and open courtyards are more significant than the building itself. Contrast this with the Western use of bulldozers to make nature fit most Western buildings.

One of the best examples of the dichotomization of opposites for Westerners is in the classification of primates. Christianity's ideology put sharp boundaries between monkeys, apes, and humans, defining them as polar opposites, thus rejecting our common affinity. An early theological view of monkeys and apes was that they were stupid, unnatural, immoral monsters and devils. Western folklore of monkeys and apes follows the Christian ideas on devolution and debasement. Just as humanity was demoted from the angels, so were apes demoted from humans. Debasement is reflected in the ideas of dual creation in which the devil tries to duplicate God's power but produces debased and distorted images. The Christian folklore of Europe applied this to man and apes, horse and ass, lion and cat, sun and moon, and night and day (Janson, 1952). The *Physiologus*, possibly written as early as the second century A.D., asserts that both the ape and the ass represent the devil (Flores, 1996; Janson, 1952). Pliny the Elder in his *Historia naturalis* of 77 A.D. and, later, Isidore of Seville described apes as wild, violent, tailless humans that live in caves (Husband, 1980).

Saint Augustine's definition of humanity was reason, a gift from God, that he believed set us apart from animals (Janson, 1952). Without reason or souls, animals were not part of God's salvation, and their suffering therefore need not concern us. A definition that put us closer to angels, with reason, allowed a sharp demarcation between human and ape. For example, the most widely used encyclopedia on natural history in the Middle Ages was *De natura rerum*, written by Thomas of Cantimpre. He says that "while the outer appearance of the ape is very human, its internal structure had nothing in common with that of man; in this respect the ape resembles man less than any other animal"

(Husband, 1980; Janson, 1952). He goes on to say that the legs, feet, and heel of apes are like ours, but that the natural condition of an ape is to face the ground. "Man alone, is capable of raising his face towards the heavens so that he may perceive the source of his salvation." His stories describe the evil and stupid ways of apes. Thomas says that apes like to play with children, but that sometimes they strangle them. H. W. Janson (1952) says that this story may be derived from a tale about apes killing human babies by bathing them in boiling water.

Other tales in the Middle Ages relate to the devolution theme in which apes are debased offspring of humans because of immorality. One fable from Hans Sacks tells about the visit of Christ to the blacksmith's shop. To show gratitude for the blacksmith's hospitality, Christ takes the old and ugly wife of the blacksmith and puts her in the forge, whereupon she emerges, young and strong. After Christ leaves, the blacksmith tries to rejuvenate another old woman, who turns hideously black and shrivels like an ape. Two pregnant women, who see the blackened old woman, give birth to two apes, who escape into the forest and become the progenitors of all simians (Janson, 1952). Another fable from the thirteenth-century Rothschild Canticles relates the story of Adam, who warns his daughters not to eat certain herbs, which would cause them to conceive monsters like dog-headed men and apes with plantlike souls. In another tale, an ape imitates the hunter and puts on boots that are weighted with lead. The hunter then captures the ape, which is unable to run. The foremost scientific scholar of this period was Albertus Magnus, who wrote *De animalibus*, the "most scientific anthropology of man allowable in the Middle Ages" (Janson, 1952). The Scale of Being for Albertus was humanity at the top, the only animal perfect in mind and body; monstra (pygmies and apes) in the middle; and brutes, or ordinary animals, at the bottom. The sense organs and hands of humans may be similar to those of brutes or monstra, but ours are more perfect because we have intellect and can put them to better use. A few thirteenth-century Apocalypses depict monkeys holding owls and sometimes riding goats, a common symbol of lewdness (Yapp, 1982). All of these animals are also considered evil, and the monkey is imitating knights and hawking. The ape became an unworthy pretender to human status as well as a grotesque caricature of man.

Christianity equated sinning with certain mental disorders and with such bodily disorders as having claws, running on all fours, and being hairy. Sinners could transform into animals (Flores, 1996). For example, the devil transformed people into werewolves or wild cannibals. Deformed offspring were the result of sin in which demons in the form of animals impregnated women. Michael Camille (1992) mentions that a woman confessed to sleeping with a horned goat in Padua in 1265. Pregnant women were admonished not to look at or think about monkeys lest their imaginations affect their offspring. Monstra were an injury to God "as when a woman will give birth to a serpent or to a dog or some other thing that is totally against Nature" according to Paré's text of 1573 (Davidson, 1991). Paré goes on to describe the hideous monsters that arise from bestiality, such as a man-pig born in Brussels in 1564 with a man's face, arms and hands above the waist, and the hindquarters of a pig.

The father of primatology, Edward Tyson, wrote a treatise in 1699, in which he denied that some monstra can arise from "any real coition betwixt the two species" (Davidson, 1991). The various monstra and their production through sinning, for example, bestiality—in which peasants in some cases literally slept with their animals—all illustrate the dramatic separation between humans and animals that was so important in Christianity. In a description of apes by the Comte de Buffon in 1791, the author even denied the intermediate nature of monstra. He concluded that they were nothing but mere brutes, incapable of thought, language, and domestication and not the smartest of animals. Buffon compares apes to what he considered savage humans, such as Hottentots. Both "should be viewed together," he says, and they have, in fact, intermixed in both species. A case could be drawn that the ethnocentric attitude of Westerners toward the primitive and degenerate other races was analogous to the degenerate ape. Other early Christian writers applied the term *ape* to all the enemies of Christ: pagans, infidels, apostates, and heretics. Western civilization was not just drawing a line between apes and humans but also between other humans, in order to unjustly dominate other people.

The book *Apes and Angels*, by L. Perry Curtis (1997), explores this theme further among the Irish as depicted by Victorian cartoonists. Prior to violent Irish nationalism, English cartoonists depicted Paddy as a degenerated man and primitive peasant. After the assassinations of Lord Frederick Cavendish and his under secretary in 1882, however, Paddy was transformed into a ferocious ape with fangs, receding jaw, pointed ears, sloping forehead, bulbous mouth, and large lips. Such a depiction appears in the English humor magazine *Punch*, May 20, 1882, under the title "The Irish Frankenstein" (see Plate 1.2). Paddy was now simianized, both physically and behaviorally.

The thirteenth-century poet Richard de Fournival remarked that birds and beasts are better understood through pictures than words (Camille, 1992). In a book on medieval "marginal art" Camille says that monkeys and apes are very prevalent. Marginal art is the images on the edges of medieval buildings, sculptures, and illuminated manuscripts. In a fourteenth-century Book of Hours are three monkeys, or babuini, at the bottom of a page, who "ape" the gestures of the wise men above them. Also depicted is a "spiky-winged ape-angel" mischievously pulling on a tail and another monkey who holds up the floor, mimicking Atlas. There are endless examples of these symbolic connections. In the margins of the Lancelot Romance is a figure of a nun suckling a monkey. The meaning in this case is that a virgin has obviously sinned and given birth to a monkey. These themes are endlessly repeated in thirteenth-century marginal art with other figures such as monkeys with lances on the backs of ostrichlike birds mimicking knights, playing bagpipes, or writing texts. Apes are also shown fighting battles, preaching to congregations, lecturing to students and practicing medicine (Sprunger, 1996).

Plate 1.2. *This depiction of the evil ways of Irish "apes," entitled "The Irish Frankenstein," appeared in* Punch, *May 20, 1882.*

A Balinese/Javanese View of the Universe

As previously mentioned, Western cosmology tends to construct binary opposites, in a dualistic type of classification. Things are either deity or demon, male or female, good or bad, but not both. Early ethnographers tended to view Balinese cosmology within a Western framework (Lovric, 1986). They constructed antitheses between compass directions such as north being toward Mount Agung, the highest mountain on Bali (good, fertile, fire, gods), or south being toward the sea (evil, disease, water, demons). Using Western cosmology on the Balinese grossly oversimplified Balinese cosmology (Lovric, 1986) because there is "flow" from one opposite to the next. For the Balinese, polar opposites can exist together. Both demons and gods, for example, inhabit the sea. It is the great purifier and a source of life. The ashes of the dead are brought to the sea for further purification. To use another example, take fire and water, which the West considers to be polar opposites. Volcanoes (Mount Agung) erupt with fire, but the source of that fire is from the underworld. Vasuki, the serpent who lies under the earth, represents both fire and water (Goudrian & Hooykaas, 1971). Water also flows into the fertile rice fields from Lake Batur, located next to these volcanoes. To Balinese, fire and water are complementary, as when cremations burn the impure and water further washes away the impurities in death rituals. Fire and water can therefore be considered as one, and the gods responsible for fire and water can be considered as various aspects of unity in the cosmos (Hooykaas, 1964). Paying homage to the demonic side of the godhead is common to tantric worship (Emigh, 1984).

In contrast to the dualistic classifications and polar oppositions of Christianity of the West, Balinese/Javanese worldviews accommodate and integrate these contrasts. First, there are intergradations between polar opposites such as black/white, good/evil, and male/female. For example, the lines separating the natural from the human world are crossed to the Balinese, and not only for animals such as monkeys but also for plants. A Balinese manuscript relates that trees and shrubs walked up to the priest Mpu Kutaran and explained their medicinal purposes to him (Hobart, 1990). Plants are not just objects for medicinal purposes but also agents. Plants, animals, and humans have complementary and mutual dependence on each other. Buddhists and Shintos, for example, do not emphasize the distinctiveness between humans and other species (Kitahara-Frisch, 1991). Animals have minds and souls. Japanese primatologists perform special services to thank the souls of the dead monkeys that they studied (Asquith, 1991). Japanese define monkeys as humans minus three pieces of hair, and the monkey is the only animal who receives *san*, the term of address used for humans.

The coexistence of opposites is also not contradictory to Balinese thinking. Heaven and hell are located next to each other, and a Balinese phrase says, "Hell becomes Heaven, Heaven Hell." The Javanese/Balinese universe is one of humans locked intimately and harmoniously into place with the natural and divine world (Pigeaud, 1967). In the Balinese world, rajas, kings, and priests become gods; their kingdoms are the center of the universe. Their universes, caste system, and language reflect an ordered and ranked system. Women are symbolically laden with both creative and destructive powers, as goddess and witch, or pure and impure (Miller & Branson, 1989). When a Brahmin priest prepares holy water, he is said to concentrate his thoughts on a divine aspect of Siwa, blending maleness and femaleness, with, for example, one male breast and one female breast. This figure represents the sexual union of Siwa and his spouse. In some of the kakawins there are references to these male/female statues of deceased kings and queens in sanctuaries where their children come to communicate with them (Zoetmulder, 1974). It is also believed that liberation can be achieved by a yogin who is able to effect this intercourse within his own body (Hooykaas, 1961).

We can thus have unity in opposites. Balinese medical doctrines contain the art of healing and the craft of witches, a so-called two-in-one, or *rwa bhineda*, that is similar to tantric ideas (Lovric, 1986). One cannot have good without evil or day without night. Mark Hobart (1985), in an essay entitled "Is God Evil?", says that the Balinese laughed at him when he asked how evil could exist if God was so good, omnipotent, and omniscient. There can be no good without evil, they said. God was both good and bad.

Balinese cosmology achieves this unity by emphasizing the center as hierarchically above the other parts. The four compass directions, for example, have their synthesis in a fifth point, the center, which is a higher unity (De Josselin De Jong, 1983; Swellengrebel, 1984; Van Ossenbruggen, 1983). This theme probably goes back to some of the world's oldest texts, such as the Rigveda, a collection of 1,028 hymns that forms a creation myth on the origin of the universe. F.B.J. Kuiper (1983), a professor of Sanskrit at the University of Leiden, says that Wisnu rose from the pillar or world tree, that is, the center, that props up the world and separates the sky from the earth. Wisnu then takes three steps. His first two steps represent the dualistic universe, and the next step unites the two worlds into an all-encompassing totality. Vishnu is therefore more important than the other gods. The cosmos has borders or spheres of natural and supernatural phenomena, but they interlock. The various spheres can be divided up into equally powerful parts, and the center is the sum of the parts. As Robert Wessing (1978) puts it, "the center can stand for the whole, while it is at the same time the point of unity which holds the disparate power as it flows together." The disparate outer spheres are the receivers of this flowing power from the center, where a harmonious balance can be maintained. So while there are borders to these spheres, there also must be continuity and the lack of marked boundaries for this power to flow. The margin between structures or spheres is called *liminal*, a transitional area between spheres such as the area between life and death, sacred and secular, or between the mortal and the divine (Turner, 1967). L.E.A. Howe (1984), for example, states that Balinese "gods eventually are nothing other than deified humans." The death of a high-caste individual is called *newata*, meaning to become a god. Humans can participate in the divine by achieving *moksa* and escaping rebirth. The Brahmana are addressed by the same honorific title as gods, which is *Ida*. The elevated titles of the Satria caste translate as great gods.

A good way in which one can see the unification of opposites is to go to the wayang, the shadow plays of Bali and Java. These plays mediate between different worlds, for example, between the seen and the unseen; and they enact the Hindu cycle of creation (Hinzler, 1981). Various intermediaries affect this mediation and solve conflicts between polar opposites. One of the most important wayang figures among the Javanese is a servant god named Semar, who is described as a hermaphroditic toad (Laksono, 1986). By mediating between forces (and sexes), Semar is at the center of oppositions and makes for the maintenance of cosmic order and harmony. He is the older brother of Batara Guru, the ruler of the universe. Semar is also the guardian of humanity, and if kings or deities abuse humanity, Semar will correct the imbalance. Thus, even the masses, the ruled, have an important relationship with kings and the deities. This harmonic unity of king and commoner and god and man is denoted in the *kawula* (subject, servant, slave)-*gusti* (master, lord, god) relationship. Such oppositions are thus unified (Laksono, 1986; Moertono, 1981). The opposites servant-master, warrior-priest and commoner-king are interdependent. They need each other for power. Their differences can be unified because of this mutuality. All things, even things as different as gods and people, are aspects of the One God. Soemarsaid Moertono (1981), for example, says that the king and the masses are similar. Each needs and protects the other, just as a jewel-encrusted scabbard needs the blade of a *keris* (dagger). Such interdependent relationships are also discussed in Bali, for example, in the Ramayana and the Babad Buleleng (Worsley, 1984).

The Balinese wayang equivalents of the Javanese servant god are Twalen and Mredah, a father and son of divine descent who may have been pre-Hindu deities and are now clowns and servants of the gods. They are the most popular heroes of Balinese wayang and perform as intermediaries between traditional and modern life, especially as translators. Twalen existed before the earth, and before gods and men. He brings about harmony, and his voice is as soft and as winning as that of a woman. His use of mantra and black magic cures the sick (Hooykaas, 1970). He is the obese, black-skinned Balinese counterpart of Semar, who is the younger brother of Siwa. Interestingly, in some parts of Bali, even Hanuman has a clown attendant. These intermediaries constantly give the Balinese a moral and spiritual worldview, especially because they are the only wayang puppets that can move their mouths. Such figures preserve and speak the past and relate to the present concerns of ordinary people.

There are many Balinese folktales with their lessons on morality and religion that have monkeys in them. Balinese animal stories are often collectively called satwa tales. The word *satwa* comes from the Sanskrit word *sattva*, or animal. Pictorial representations of these oral tales are also carved into temple walls. T. M. Hunter (1988) suggests that the pictures of animal tales found in the lower tiers of temples impart religious doctrine to the masses, whereas the elite was educated from the great Sanskrit epics. The scenes from the latter stories are found at higher levels of the temples.

In many stories the gods transform themselves into animals. These incarnations are called *avatara*. Wisnu's and Brahma's incarnations, for example, are such animals as a boar, a tortoise, a man-lion. There is a Hindu story in which Brahma and Wisnu race each other to get to the top of a monument (Sukawati, 1979). Brahma changes his form to that of a bird, whereas Wisnu changes to a pig. The bird, however, cannot find the top, and the pig cannot find the bottom. Note that in this story, the top of the monument can also be the bottom, according to one's perspective. The pig, however, does encounter the earth goddess, and he catches her, makes love to her, and impregnates her. When the goddess cries, Wisnu changes his form and tells her that he is not really a pig but the god Wisnu. He tells her that the child will be a powerful ruler of a large country. Both Brahma and Wisnu realize the quest is futile and go to Siwa to apologize. In contrast to this story and the others involving Hanuman, for example, Christian worldviews attributed demonic evil characteristics to transformations, which were either a product of sin or of individuals who were really witches, transforming themselves into animals (Rohrich, 1971). These "tribal" folktales, according to Lutz Rohrich, are in marked contrast with European folktales that depict animals in their place as animals because European civilization has "ousted nature."

Monkeys and other animals are often depicted as benefactors to humans; if not, then they are punished. In the tale of the goldsmith, a tiger and monkey befriend a falsely accused Brahman (Hunter, 1988). A Brahman rescues a monkey, a snake, and a tiger, which fall into a well. The animals warn him not to pull out the goldsmith, who is also in the well. In an act of charity, the priest rescues the man, who falsely accuses the priest of killing the king's oldest son. The priest is imprisoned, and in retaliation the snake bites the king's younger son. One of the morals of this story is that "it is more dangerous to befriend evil-doers than wild-animals" (Hunter, 1988).

There are many other such stories in which a monkey saves a princess from a bull (Hooykaas, 1963) or monkeys bring fruits to a banished sister, who eventually marries a prince (Bagus, 1968). Another tale is about a female monkey who helps a hunter climb a tree to escape from a tiger (Marrison, pers. com.). The hunter repays her by killing her and her family, but she goes to heaven, whereas he goes to hell. Balinese who are violent toward animals find a hell awaiting them in the form of being trussed and roasted over a fire like a pig (Hunter, 1988). Another story tells how a boy traps a number of animals, including a monkey (Alibasah, 1990). The animals promise to help him if he lets them go. They all eventually do help him, so that the boy ends up marrying a princess. The monkey helps by changing into a man and riding a horse that has changed from a deer and winning a horse race. Yet another story depicts a monkey as a heavenly musician.

A story translated into German by Jacoba Hooykaas (1963) is "Boosard, der Affenkönig." The leader of the monkeys, named Boosard, torments a young girl because she won't marry him. When her father goes into the fields to work, Boosard keeps taking things out of the house, including a pig. Her father, Cubling, concocts a plan wherein he pretends to be dead and lies in the bier. The girl agrees to marry the monkey king if he will help dig a grave and bring back all the stolen property. All the monkeys help in digging a grave, weaving a mat, and boiling water. Only two monkeys clean up the place, a male and female, and they detect that Cubling is not dead and run off. When the hole is deep enough and all the other monkeys are in it, Cubling and his daughter kill them with boiling water and bury them. The remaining two monkeys reproduce, and that is how they continue to exist up to the present day.

There is a Balinese folktale in which a tiger makes the statement "You human beings are said to be so superior to us animals, but Man does not repay the help or service of other creatures" (Alibasah, 1990). The kind boy, who had just let the tiger out of a trap, denies the tiger's statement. The tiger then says that he will not eat the boy if they can find another animal that says that man repays the kindness of other creatures. They meet a horse, an ox, and a hawk, who say that man only beats them or shoots them and that the tiger should eat the boy. The mouse deer, however, tells the tiger to get back into the cage. Once back in the cage, the mouse deer tells the boy to go home and that one should help some, but not all creatures.

Countless tales exist in Indonesia relating how smart monkeys are (Alibasah, 1981). For example, in "Crocodile and Monkey" a monkey outsmarts a crocodile that wants to eat him. The crocodile is upset that the monkey is eating all the fruits that the crocodile wants for himself. He asks the monkey to climb on his back to visit his sick father. In the middle of the river, he then informs the monkey that his father needs a monkey heart in order to be cured. The monkey says he needs to return to shore to get his heart because he left it hanging in a tree. Once on shore, the monkey runs up the tree and calls the crocodile a stupid fool for spitting out food in his mouth. This story is carved on a temple near Borobudur in Java (Sankaranarayanan, 1992).

Not all of the stories about monkeys are ancient. One Indonesian story by Prijana Winduwinata (1978) is called "Don't Become a Teacher." It is about an idealistic and suffering monkey who becomes a teacher in order to build the new country's future.

Balinese Monkeys: Animals or Gods?

What is the Balinese worldview of monkeys? In contrast to Western worldviews depicting monkeys as disgusting and stupid sinners, Balinese views are much more complicated. Monkeys are sometimes disgusting and stupid to the Balinese, but they are also smart, loyal, and brave. A few examples from the sacred Ramayana Kakawin and various folktales should suffice to explain this view.

In the Ramayana Kakawin monkeys are invaluable in the fight against evil and demons. Rama calls the monkey king Sugriwa his beloved friend who helped him get Sita back and win the war. The powerful monkey king Bali and brother of Sugriwa, however, is a good friend of the demon Rawana.

Another important monkey is Hanuman, the monkey chief. The name Hanuman derives from the Tamil word, an-manti, meaning "male monkey"

(Bhattacharji, 1970). As the leader of the messengers, Hanuman is ordered by Sugriwa to find Rama. Rama calls Hanuman a good and excellent messenger after Hanuman finds Sita and brings a letter from her back to Rama. Hanuman is the ideal messenger because he is at the borders of the animal world, the human world, and the world of the gods.

Monkeys have animal, human, and godlike qualities. They are noisy and mischievous, but they are not ordinary animals. The gods are tricking the demon Rawana by sending a human with a monkey army. When Rama bemoans the fact that he is unable to kill Rawana, he tells Sugriwa to go home. The latter says that if they do go home, then "all people observing us will humiliate us. Ah! Monkeys. They are really debased animals who have no regard for the world. So they would say."

Monkeys and Hanuman have some humanlike qualities. Sugriwa is a powerful, noble, and good-hearted monkey king. Rama appoints Sugriwa's nephew Angada a crown prince. Hanuman especially has meritorious characteristics. He is an herbal healer and a very wise, devoted and altruistic servant of Rama. Rama addresses him as a holy man, conversant in scriptures and formulas, when he first talks with Hanuman. Some of the *sadwargga*, or six enemies of the body, do not tempt monkeys. For example, the Ramayana Kakawin says, "You cannot give jewels, gold or pretty clothes to monkeys to bribe them. All the valuables have no value to them, what they want is only fruit, nothing else. Monkey heroes are very loyal." They have no love of luxury, splendor, or worldly goods. Soewito Santoso (1980b) says that Hanuman can appear in different shapes and numbers because he has the Javanese power of the division of personality.

The Uttarakanda has more on Hanuman and Bali. The Uttarakanda was written in the tenth century, and it is the last of seven parts of the epic poem, the Mahabharata. One of the sections describes Hanuman's childhood (Worsley, 1984). He saves the sun from the demon that eats it and the moon during eclipses. Hanuman has the gift of invulnerability from his father, the wind god Bayu (from the Sanskrit word *bayu*, meaning wind), and the gift of enlightenment from Surya, the sun god. Under Surya's tutelage, he becomes unequaled in all the sciences and ascetics. He is also the ninth author of grammar. The learned monkey chief also described the four different ages that are important in Hindu and Javanese/Balinese cosmology (Dowson, 1972; Laksono, 1986). An interesting part of Hanuman's story in the Uttarakanda is that a heavenly sage curses Hanuman for ruining hermitages. His curse is that Hanuman cannot have consciousness of his supernatural physical and spiritual strength.

In another kakawin is a story in which the gods of the four corners of the world turn into animals (Zoetmulder, 1974). Frightened at the arrival of a demon, Indra changes himself into a peacock, Yama into a crow, Baruna into a goose, and Dhaneswara into a chameleon. Brahma grants the demon immortality, but when he ridicules Siwa's guardian for having the head of a monkey, Siwa curses the demon and says that monkeys shall destroy his residence, kill his kin, and that Wisnu in human form will kill him. This story is a marked contrast with the Western stories that commonly ridicule monkeys.

Do the Balinese have a monkey god? Monkey gods are present in several Asian countries, in India, China, and Japan. The monkey god in India is, of course, Hanuman. The monkey usually associated with Hanuman is a langur belonging to the genus *Presbytis* and species *entellus*, but the privileges of deification appear to apply to all monkeys in India, such as the rhesus macaque (Southwick & Farooq Siddiqi, 1985). The monkey deity in Japan is called Saruta Biko. He is seen not only as a messenger from the mountain deity but also as a mediator between deities and humans and between heaven and earth (Ohnuki-Tierney, 1987). In Japan, the monkey was a shaman whose curing powers stemmed from the mountain deity. In Bali, however, the situation is more complicated, although Fred B. Eiseman, Jr., and Margaret Eiseman (1988) as well as Hunter (1988), call Hanuman the monkey god. Hanuman, Bali, and Sugriwa, the monkey king, do have godlike mystical powers. There are generally eight supernatural powers, called *astaiswarya*. These powers are said to manifest themselves when the yogi's concentration becomes most intense, burning away his past sins. The ability to fly and to levitate are two of the eight Balinese mystical powers (Hobart, 1985), and in the Ramayana Kakawin, Hanuman and other monkeys fly so swiftly that trees are blown down. Not even Garuda and Surya, the sun god, and the wind are a match for Hanuman. He flies above the planets, the sun and moon. Only Hanuman, along with the wind god and Garuda, can cross the ocean.

Like Hanuman, Bhima, the warrior-priest, is also the son of the wind god. Bhima and Hanuman are half-brothers. Adrian Vickers (1989) refers to Bhima and Hanuman as the only allies of mankind, in part because they both share in the terrifying power of demons and witches. Hanuman is more courageous than the deities, says the Ramayana. Hanuman has supernatural strength, but also other powers, such as the two mystical powers of enlargement (*mahiman*) or diminution (*animan*) (Hobart, 1985; Hooykaas, 1964). Arrows cannot hurt him. Hanuman's merits are perfect and pure, to be taken as an example by those ruling over the people, states the Ramayana. These eight powers were attributed to Dwijendra, also known as Dang Hyang/Mpu Nirartha, "The Unworldly" (Rubinstein, 1991). All Balinese Brahmana claim descent from this legendary Javanese priest, who is an earthly descendant of the god Brahma. The manuscripts state that when Dwijendra flees from Java to Bali, he is promptly guided or given directions by a monkey (Rubinstein, 1991). The same story also describes the origin of the vow that Brahmana may not harm monkeys. Hedi Hinzler (1974) says that there are royal families who consider the god Subali of Talankir as their founding ancestor and that there is also a reference to "a not yet traced King Subali" according to several *Babads* (documents that bind the living to the dead). Such depictions of a monkey god in Ksatriyan (noble) ancestry or as a king is especially germane to calling the Ramayana monkey kings, gods in Bali because of the veneration of ancestors and the care of their souls in rituals. R. Friederich (1959) mentions a manuscript called *kapiparva* that contains the history of Sugriwa, Hanuman, and other monkeys, but no one else seems to mention it.

There are many depictions of the Ramayana monkeys, especially in paintings (Worsley, 1984). T. M. Hunter (1988) describes the appearance of Hanuman in a Balinese wayang painting known as *The Baskett Tabing*, where the "heroic monkey-god is surrounded by a flame-encircled inset appropriate in depictions of divine beings." The inset is commonly called an aureole, which typically surrounds kings, ascetics, or gods. It may also indicate divine approval and protection to achieve their destiny. The radiant glow occurs when the soul is released through the fontanelle and unites with the Supreme Being during yoga. The scene is from the Ramayana, and Hanuman has just returned Sita to "an amazed and visibly gratified Rama" (Worsley, 1972). A painting of Hanuman wearing his *kain poleng* (checkered cloth worn by temple officiants and giving magical protection in battle; Gralapp, 1967), with Rama and Sita is at the Batur Temple on the side toward the lake (see Plate 1.3). There are also statues of Subali and Sugriwa. Prince Sukawati (1979) mentions that after he had finished painting the Goddess of Batur with gold leaf, the head of all of the *pemangkus* (commoner priests) in Batur asked him to paint Subali and Sugriwa in gold leaf. An object depicted in Helen I. Jessup's book (1990) is a kris holder from Lombok that shows Hanuman with "gold-edged loincloth, gold

Plate 1.3. *A painting in the Temple at Batur showing Hanuman from the Ramayana.*

necklace, sumping or head ornaments, shoulder ornaments, bracelets and anklets" that indicate his high rank.

One of the most interesting depictions of Hanuman and monkeys occurs in a nineteenth-century painting from Kamasan (Worsley, 1984). It is a visual presentation of a section of the Ramayana, specifically that part when the monkey army built a causeway to Lengka. Monkeys, as animals, are visible at the bottom of the painting. They are hard at work, laboring to restore harmony to the world. This painting is a moral discourse on kingship, and Worsley (1984) compares these monkeys to the natural world of commoners in service to the king. At the bottom of the painting is a frieze showing monkeys instructing a tiger and two bulls on some moral point. Located above these monkeys on the painting are more monkeys or *peluarga* (hybrids of monkeys and humans or some other animal), as Anthony Forge (1978) calls them. Dressed as Ksatriyan nobles like Sugriwa, peluarga are under the command of Sugriwa and serve Rama, shown wearing a crown. This interpretation is congruent with the depictions of Twalen and Mredah, who symbolize the harmonic unity of commoner/king and gods/humans by wearing kain poleng. The most interesting figure in this painting is Hanuman, shown in white, wearing his kain poleng and donning a *supit urang* coiffure. Young aristocrats use such a hairstyle. He is flying over the causeway and the ocean among seven priests and is grasping for the sun. Hanuman is, of course, flying to Lengka in his quest for Sita, but as Worsley (1984) points out, Hanuman is closer to the sun than the priests are, thus indicating his priestly role and ritual labor. Hanuman is about to grasp the sun and, in fact, to become one with the sun deity, just as a siwaist priest does through meditation and ritual. The souls of dead priests are also said to merge with the sun. Hanuman's energetic leap shows the passive priests what they should be doing to help restore order to the world, and perhaps also, as Peter J. Worsley says, to usurp their power as a rival to the seven priests. Hanuman's portrait recalls his story in the Uttarakanda (Worsley, 1984). Both Hanuman and Garuda are also symbols of the enmity between monkeys and the nagas (snakes with dragonlike heads). Numerous woodcarvings attest to this theme, showing Hanuman astride twin nagas. Other folk carvings show monkeys accompanying fishermen. Monkeys are commonly known to warn of and to attack snakes. Both Garuda and Hanuman are associated with the god Wisnu. Worsley (1984) maintains that peluarga, Hanuman and Nala are symbols of cosmic energy for commoners.

The Liminality of Monkeys

Robert Wessing (1978; 1988) wrote about the importance of the connections between Javanese cosmology and the environment with specific reference to the symbolic elaboration of the tiger. It is surprising that the monkey has not received similar treatment. Like the tiger, the monkey is an animal. Both live in mountainous forests. They are attracted to the human world; their ecological niches overlap. While the tiger eats the animals attracted to the gardens of farmers, including sometimes the farmers themselves, the monkeys eat the crops. There is, for example, a Javanese drama in which a monkey is killed for being one of the six enemies of the rice harvest (Rassers, 1925).

Monkeys have other ambiguities. They are often found in so-called marginal areas where many demons live, such as ravines, graveyards, slopes, and hollows. It is at the intersection of these borders where danger is present, for example, in the marginal area where land and water meet. The souls of humans who have died bad deaths live in ravines and attack people there (Hobart, 1985). These are the areas where Siwa and Uma metamorphasize into their demonic forms such as Durga, the chief *leyak* (witch/sorcerer) or Rangda, the arch-witch of Balinese mythology (Howe, 1984; see Plate 1.4). These demons are responsible for droughts and epidemics. Witches often transform into pigs and monkeys to enact evil. The head of the Monkey Forest Managerial Committee in Padangtegal probably had this in mind when he apologized to a tourist, who had just broken his back, by saying that the temple committee had not made an offering to Rangda, (Lattin, 1994). The tourist had approached a monkey next to a stone carving of Rangda at the Pura Dalem (Death Temple) when the monkey "suddenly bared its fangs, thrust out its tongue, and lashed out" (see Plate 1.5). This caused the tourist to jump back and fall when the board he was standing on collapsed. Previous to this incident, another tourist went crazy after climbing on the wall of the Pura Dalem. Again Don Lattin (1994) quotes the committee head as saying, "We should have made an offering to Rangda right away."

In another example, Prince Sukawati (1979) describes how one of his aunts practiced black magic and transformed herself into a "black magic monkey." Monkeys, then, are also potentially demonic, and their animalistic characteristics are a source of disgust for Balinese (Foster, 1979; Geertz & Geertz, 1975). One interesting demonstration of this aversion was the curious lack of monkeys represented in carvings or silver and other artistic objects sold in the hundreds of shops near the Monkey Forest. Assisted by Earthwatch volunteers, we quantified the representations of monkeys sold in sixty shops. It is common knowledge to the shopkeepers that tourists came to the Monkey Forest to see monkeys, but only a quarter of the shops sold more than a handful among hundreds of other animals such as frogs, fish, birds, geckos, dolphins, and whales in their inventory. Half of them did not sell a single object depicting a monkey! It must be emphasized that although Durga is a dangerous enemy, she is also a powerful and helpful patron.

One of the holy men of the village of Ubud is a *pemangku dasar:* a commoner priest, not a Brahmana. He is a fifth-generation priest (from pre-Hindu ancestors) who had a vision of Hanuman when he was a boy. This vision was an omen to serve god as a pemangku. Hanuman is the symbol for his energy, Batara Bayu, that he also called Sang Hyang Murti and Bima Sena, and he has a picture of Hanuman with white fur in his house. Hanuman is sometimes called the white monkey. White is said to be the color of the east. The priest also always dresses in all white with a white turban, as do most priests. This priest has a special duty to give offerings to Hanuman at the Monkey Forest on full

Plate 1.4. *This statue of Rangda devouring the dead appears in the outer courtyard of the Pura Dalem.*

and new moons so that the monkeys are not dangerous to people. The offerings on the full moon would presumably be to Hanuman, the monkey god, whereas offerings on the new moon would be to the demonic side of monkeys. The offerings include such things as corn and follow the Hindu *saka* calendar, which is important in agriculture and other festivals (Covarrubias, 1937). This priest has other visions as well. Sometimes he sees monkeys as tigers or cobras. The

Plate 1.5. *This adult male exhibits his formidable canines and tension in this yawn.*

Balinese tiger is now extinct, but it is a common and powerful symbol throughout Southeast Asia (Wessing, 1988). It represents an ancestor, a shaman or royalty. When I asked the priest if he ever saw monkeys as humans, he was disgusted and said never.

The pemangku assists at the Pura Dalem, and he performs marriages and other village ceremonies, including the twelve-day-since-birth ceremony, or *ngelepas hawon*, for the community. The pemangku is also a *balian* (healer) who goes into trance and communicates with the spirit world to inform the parents on the ancestral reincarnation in the infant. This priest is highly regarded in the village by commoners, especially, it seemed to me, by the working-class people. Upper-class individuals spoke of him as unmodern and uneducated. One day the pemangku invited me to walk down the street with him to the Monkey Forest. I was very tired at the time and almost declined. The experience was amazing, as all the people who were "away" (Bateson & Mead, 1942) and never previously acknowledged my existence were looking at the two of us very deferentially and addressing him with the utmost respect. It was as if a whole other part of Bali was opened up to me. It is therefore appropriate that this pemangku is also a liminal figure and mediator between the spirit world and the human world, between animals and humans and between the living and the dead. This pemangku and the entities that he addresses are also liminal (Wessing, 1978). He obtains *sakti* (magical power) from these worlds to help his community (Geertz, 1994).

A number of different bayu (the vital breath of all life) refer to the monkeys of the Ramayana and not just to Hanuman (Hooykaas, 1978). For example, Anala and Sugriwa are also mentioned in Babayon mantras and formulas. There is a mantra for Brahman priests called Hanuman-Kavaca (Goudriaan & Hooykaas, 1971). The god Bayu is also associated with seven other deities that are in the body of a king, according to the Serat Rama (Moertono, 1981). The king should therefore also have Bayu's qualities, which are keen awareness and deep insight. Because Hanuman has the ability to fly, he and others such as Angada, the son of Subali, are invoked in mantras and drawings to help clear the sky (Hooykaas, 1970). Such actions are necessary in Balinese shadow theater, and no dalang (puppeteer) can start or continue a performance in the rain. Note that Bayu, the wind of life, can easily communicate between the various worlds. The west is the region of the dead and is associated with the underworld in a vertical three-part division with the heavens represented by the east and north (Swellengrebel, 1984). The dalang is not just a puppeteer. He begins a performance with priestly activities and becomes divine and very similar to a Brahman priest, according to I Gusti Bagus Sugriwa (cited in Hooykaas, 1970). The dalang attracts benevolent spirits and chases away malevolent ones. He therefore acts as an exorcist priest and averts danger that threatens people. It is interesting that the wayang performances that take place at the third-month festival (*nigang sasihin*) sometimes include stories of Hanuman's birth and the fight between Subali and Sugriwa (Hinzler, 1981).

In many respects it is the admirable qualities of Hanuman, such as his strength and bravery in battle, that disqualify him from being a god. In Howe's (1984) pantheon, the gods represent meritorious thoughts, devoid of anger, hate, envy, and jealousy, but they are abstract and "lack an action component." "Gods do not do anything qua gods," says L.E.A. Howe (1984). The monkey is therefore in a border area between the animal/demonic world, the world of humans and the world of gods. The monkey has power to transform between these worlds and to transfer power between these worlds, and that is why the monkey is both feared and revered and why Hanuman is Rama's ideal messenger. A monkey is an unclassifiable animal and a transitional one. In Victor Turner's (1967) usage, transitional beings are particularly polluting since they are neither one thing nor the other. Hanuman's divinely endowed spiritual and physical strength "enable him to transcend the normal barriers between refined gods and humanity on the one hand and coarse demons and beasts on the other" (Jessup, 1990). This transcendence is shown in Helen Jessup (1990) in a painting from Kamasan, a village near Gelgel, Bali. Hanuman is shown in white, wearing a magical black and white cloth, and his central position in the painting shows him "mediating" and "reconciling the disparate elements."

The human/godlike qualities of Hanuman and the ambiguity of monkeys also force the Balinese to incorporate them into discussions regarding their society. On the one hand, according to Turner's (1967) analysis, how might these qualities of monkeys represent aspects of Balinese culture; and, on the other hand, what aspects of the monkey can be gained by, for example, the study of the habits of monkeys? The whole relation between humans and monkeys is continually open to hypothesis. In other words, the Balinese use monkeys to deliberate about themselves, as do the Japanese (Ohnuki-Tierney, 1990). Recall, for example, the peluarga (half-human, half-monkey) in wayang paintings that come from the Ramayana story (Forge, 1978). These creatures assist Rama in his quest for Sita, and they are intermediaries between the upper and lower worlds. Peluarga are also intermediaries between the Hindu gentry and commoners. In chapter 3, I also relate a story told to me about the effects of tourism on monkeys and how this is analagous to the effects of tourism on the Balinese.

It is appropriate to end this chapter by returning to some of the previous scientific names given to the monkey species of central interest in this book. The scientific name for the crab-eating or long-tailed macaque is presently *Macaca fascicularis*, but a previous name for it was *Simia cynomolgos*. The name, *Simia cynomolgos*, comes from Johann Schreber, who used it in 1774 (cited in Napier & Groves, 1983). Solinus used the genus name, *simia*, or *simius*, in the fourteenth century, meaning an ape [*sic*] that liked to imitate men. The name derives from *similitudo*. According to Janson (1952), Solinus remarked that hunters caught apes after they blinded themselves by incorrectly imitating the hunters who pretended to smear birdlime in their eyes. Note that it is always the feminine form, *simia*, and not the masculine form *simius* that is used because of the former word's association with the debased Eve. In a similar manner, no pun intended, the monkey, *le singe*, is an anagram for *le signe*, the sign which points to "the danger of mimesis or illusion in God's created scheme of things" (Camille, 1992).

Another species name formerly used for this monkey was *irus*. W.C.O. Hill (1974) and many other researchers used it, deriving from Georges Cuvier in 1818. Cuvier chose the name *irus* because it is the name of a notorious beggar, a vagabond, constantly eating, drinking, and even insulting and fighting Odysseus in Homer's Odyssey (Hill, 1974; Rees, 1960). Cuvier's attitude toward this monkey is thus very much in line with European attitudes of his time. In 1757, for example, Osbeck said that it resembled "all others of this genus in dirtyness, lasciviousness, drollery" (cited in Napier & Groves, 1983).

There is, however, a further twist to the name Irus. The real name of Irus is Arnaeus (Rees, 1960). The nickname Irus was given to him because he ran errands for others. The name Irus thus appears to derive from Iris, who was the divine messenger of Zeus. Iris is a Greek goddess mentioned in Homer's *Iliad*. In R. Fugles's (1990) translation, Iris appears in human form as "the wind-quick messenger." She was a sister to the winds; grandaughter of the sun; the rainbow; and harbinger of fair weather (Gordon, 1943). She also conducts, as a charioteer, the wounded Aphrodite to Olympus. Iris is an intermediary between humans and Zeus because the latter does not communicate directly with humans. With this ironic twist, we come full circle. The old species name of this monkey derives from the Greek goddess and resembles Hanuman in the sense that both are divine messengers and swift as the wind. Likewise, the early Greeks interpreted many of the creatures mentioned in the Indian epics as real and disgusting monstra (Wittkower, 1942). The Greek epics *The Iliad* and *The Odyssey* resemble the Indian epics in many ways, including the times when they were written. So the Balinese dialectical view of monkeys is instructive to us after all. The next chapter explores one of the reasons that the Balinese were not as misled as the Europeans for so long. The Balinese, in contrast to Europeans, have a long commensal relationship with this monkey species.

Chapter 2

Primate Commensalism

Human Influences on *Macaca fascicularis*

Up to now, I have discussed the long Asian or Balinese cultural history of interactions between humans and other primates, such as monkeys. The general topic of this approach is called commensalism. The term *commensal* is commonly defined as "(1) one who eats at the same table, (2) an animal or plant living with another for support, or sometimes for mutual advantage, but not as a parasite" (Southwick & Siddiqi, 1994; *Webster's 20th Century Unabridged Dictionary*). Primate commensalism is an inadequately known and recognized area by primatologists. In a publication of a recent symposium of the International Primatological Society (Gautier & Biquand, 1994), the more common type of commensalism is one in which nonhuman primates take advantage of human food, such as crops or dump sites, where they are often uninvited dinner guests. The impetus for primate commensalism as a new topic of study for primatologists appears to derive from increasing conflict between monkeys and humans as the latter encroach upon the former, forcing animals to become, if they can, commensal with people. Many primatologists treat this commensalism as "new," while missing the long-term relationships between some of these primates and the people who live near or with them.

Two of the most widely distributed nonhuman primates in the world are (1) the rhesus macaque, *Macaca mulatta*, followed by (2) the long-tailed macaque, *Macaca fascicularis*. Both of these species are closely related, belonging to the same species group (Fooden, 1991a; Hoelzer & Melnick, 1996). The rhesus monkey is found from Pakistan and Afghanistan in the west to China and Vietnam in the east. Ecologically, the species is equally diverse, being found in tropical as well as temperate habitats (Southwick, Yongzu, Haisheng, Zhenhe, & Wenyuan, 1996). The long-tailed macaque is found from Bangladesh to Vietnam and east to the Philippines and Timor. Its habitat is basically tropical, from the seashore to the montane forests and to habitats shared with humans in their villages, towns, and temples throughout Southeast Asia (Fooden, 1991b; 1995; Hill, 1974; Pocock, 1939; Wheatley, 1980b). With such a diversity of habitats and widespread distribution throughout the world's largest archipelago, it is not surprising that the long-tailed macaque is also one of the most morphologically diverse. J. R. and P. H. Napier (1967), for example, list more subspecies for *fascicularis* than for any other species of macaque. *M. f. fascicularis*, the most common subspecies, has had fifteen different names assigned to four different genera (Chasen, 1940).

Rhesus monkeys are considered to be one of the world's best examples of primate commensalism (Southwick, Siddiqi, Farooqui, & Pal, 1976; Southwick & Siddiqi, 1994). In a comparison of population surveys in north central India over a 30-year time interval, Southwick and Siddiqi (1994) found the same high percentages (86 to 88 percent) of the population in commensal or semicommensal habitats. In fact, rhesus monkeys were becoming more commensal by living closer to people in the villages, towns, and temples and were decreasing in semicommensal habitats such as along roadsides and in parks. Only 12 to 14 percent of the populations sampled lived in forests. Comparisons of rhesus monkeys in India with macaques in Bali are interesting in that both species are often considered to be notorious pests but at the same time sometimes protected by the Hindu attitudes of the local people (Blanford, 1887; Southwick & Siddiqi, 1985). Igbal Malik (1988) found such traditional attitudes to be effective in protecting rhesus macaques at the ancient city site of Tughlagabad near New Delhi. There are several other macaque species besides *M. fascicularis* and *M. mulatta* that thrive in habitats disturbed by humans.

A. F. Richard, S. J. Goldstein, and R. E. Dewar (1989) have identified *M. radiata* and *M. sinica* in this same category, and they call all four of these macaques weed species. Is it, perhaps, not accidental that long-tailed monkeys share at least the same degree of commensalism with humans? *Macaca fascicularis* is often cited by early explorers as being closely associated with humans (Blanford, 1887). I can also personally attest to their raiding gardens, including mine in Borneo. What are the depths of this commensalism, and what other types of interactions and relationships between humans and this species of macaque are there?

It is noteworthy that where the geographic distribution of the rhesus monkey stops in the Indochina area, the distribution of the long-tailed macaque begins. I have proposed elsewhere that malaria may delimit the ranges of these macaques and that the long-tailed macaque's genetic immunity to this disease helped facilitate its divergence from an ancestral macaque and its dispersal into Southeast Asia where malaria is rampant (Wheatley, 1980b).

There are many explanations for neglecting to account for primate commensalism. The main reason for this bias is probably ideological. As previously explained in chapter 1, Westerners have a long tradition of not considering humans as part of nature. Many primatologists, therefore, seek out primates for study in habitats that are "undisturbed" by people. There are certainly good reasons for this. Some primates cannot adapt to habitats that are overly modified or destroyed by people. One may also want to know how through their evolutionary past some primates have adapted to their environment in the absence of human interference. Some of these environments are rapidly disappearing, and we may never know their methods of foraging, predator avoidance or interspecific relationships unless we do such studies. By deliberately seeking such "undisturbed" areas, however, we should not ignore the possibility that human-nonhuman primate interactions might actually be natural. Humans affect even so-called "undisturbed" areas. As Peter Vitousek, Harold Mooney, Jane Lubchenco, and Jerry Melillo (1997) note, "no ecosystem on Earth's surface is free of pervasive human influence." Not only is human modification of habitats substantial, but it is also growing. It is certainly ironic that human help is increasingly required to save wild species.

Accepting our primate evolutionary past is perhaps not as difficult for anthropologists as for others, of course, but expressing exactly what and how primates can tell us about ourselves is difficult even for anthropologists. For example, how many primatologists are happy about the treatment of primatological or even paleoanthropological topics in introductory textbooks? Not many, I suspect. Are the language experiments on chimpanzees adequately discussed? Are anthropologists even receptive to the idea of discussing culture in chimpanzees or fossil humans? We need more integration of the advances from all four or five of the subfields.

Some other explanations for the bias against studying human influences is that many primatologists are not equipped to study people. This is especially the case if they are not anthropologists and don't know the local language or culture. Whatever the reason, and I risk offending everyone with such generalizations, it is important to realize that we have neglected the role of humans in primatology. The important role local people have on conservation would be missed. The role we as anthropologists can play in conservation efforts would be absent too.

I would now like to consider topics from my research on *M. fascicularis* that point out the deficiencies of our research when human influences are not considered. The first example is from Indonesian Borneo. Here I point out that even when we think we are studying primates in undisturbed habitats, this is far from true. I then discuss the antiquity of primate commensalism. The next example is from the Island of Ngeaur in the Republic of Palau where I examine the effects of human hunting on the behavior of these monkeys. It is important to note, too, that on this island there is no tradition of primate commensalism. The Germans introduced these animals between 1909 and 1914, and the Ngeaurese say that the monkeys should be eradicated. Last, I briefly examine Balinese temple monkeys to reveal their dietary dependence on humans. In addition, I discuss the behaviors of these animals in regard to cultural behavior.

Indonesian Borneo

In pursuit of my doctoral degree, I followed my advisors, Peter Rodman and Donald Lindburg, by going to the Kutai Nature Reserve in Kalimantan, Indonesia. I spent twenty months studying a troop of these macaques. Part of the rationale for my work was to study the ecological effects on the social behavior of *M. fascicularis* in an undisturbed habitat. This "undisturbed" habitat was tropical lowland evergreen rain forest. Southeast Asia was estimated to have one–fifth of the world's evergreen rain forest, and Indonesia was estimated to have two–thirds of the estimated 240 million hectares in Southeast Asia (Walker, 1980). In the 1960s and 1970s, much of our work on primates was on species in habitats providing optimal observation conditions. Such habitats were, for example, on the African savanna, where Sherwood Washburn and Irven De Vore (1961) did their classic studies on baboons. Results of these and other studies provided us with a number of hypotheses regarding the evolution of social behavior and ecology. For example, the large multimale troops of baboons were seen as specialized adaptations to terrestrial foraging, and large multimale troops of arboreal primates were seen as atypical. Most primate species, however, are arboreal rain forest animals, and if our models on primate social behavior were to be complete, then more studies on arboreal rain forest animals would be needed. My work in Kalimantan was to help realize just such a goal.

These macaques were certainly arboreal (see Plate 2.1). They spent more than 97 percent of their time in the trees, but imagine my surprise when I realized that the core area of these arboreal macaques was in "disturbed" areas. These areas were secondary forest, especially along the rivers and streams in the study area. The most common cause of secondary forest even today (Whitmore, 1975) is agricultural, especially *ladang*, or slash-and-burn, agriculture. According to the local people, there were such gardens in the macaques' core area between thirty and fifty years ago. This type of agriculture, sometimes called swidden agriculture, consists of clearing the forests, burning the dry brush and trees, planting crops, especially dry rice, for a few years, and then abandoning or fallowing the fields for a number of years (see Plate 2.2). Farmers in Borneo tend to favor the riverine areas for planting their crops. Such areas are more accessible by boat. The deposition of silt from flooding also makes the soil more fertile and easier to work. Not all secondary forests are the products of slash-and-burn agriculture. The flooding and high winds, coupled with the shallow root systems of many rain forest trees, also account for the production of secondary forests. Both human and nonhuman produced secondary forests

Plate 2.1. *The study troop returned to this tall sleeping tree or refuge tree at dusk. I took this photo from the top of a nearby tree at my study site in Borneo.*

Plate 2.2. *Ladang, or swidden agriculture, in Indonesian Borneo.*

seem to predominate in riverine areas, exactly the habitat in which the animals spent about two-thirds of their time.

The storms can be very violent. Usually I had about two minutes' advance warning of such storms because of their horrific noise. The sound of rain beating on the foliage increases in intensity, and then trees and limbs start falling around you. One of my most vivid memories is of watching and literally feeling the roar of rainfall coming toward me down the river. I could not distinguish where the rain ended and the river began; it was all just one wall of water roaring toward me. It was impossible to stay dry under such circumstances. Hiding behind buttresses of huge trees worked for a few seconds, but the rain came from many different directions at once. Wearing a poncho did not help either, because the rain just beat through it or with the humidity and heat so high, I got soaking wet from condensation before the storm even hit.

By recording the location of the troop on a map as the monkeys foraged throughout their home range, I determined their core area in which they spent three-quarters of their time. I next chose one of these twenty-seven hectares at random for intensive analysis. The sampled hectare was fifty meters from a river. It contained 1,179 trees and lianas, most of which were small, that is, between 4.5 cm. and 12.6 cm. in diameter at chest height. Calculation of the basal areas of all trees in this 100-meter-by-100-meter plot showed that certain pioneer trees typical of secondary forests dominated the hectare, such as *Macaranga pruinosa* and *Callicarpa farinosa*. An analysis of the vertical structure of this hectare also showed the dominant position of these trees in the canopy. There were only three stories. The top story consisted of one emergent fig tree about forty-five meters tall, and the second story was dominated by the *Macaranga* and *Callicarpa* trees at about the twenty-five meter level. These features of this sample hectare are characteristic of secondary forests (Richards, 1966; Whitmore, 1975). There was also evidence of hand logging on ironwood trees (*Eusideroxylon zwageri*). More information on these and other results can be found in Bruce Wheatley (1978b; 1980b) and Wheatley, Harya Putra, and Mary Gonder (1996).

One of the important features of pioneer tree species is that they fruit several times each year. These trees are present in high densities, and their fruits are rapidly renewing. About half of the 1,179 trees and lianas in the sample hectare were edible. The animals would break up into small groups and forage, often individually, in these trees along the streamside areas. In the sample hectare, for example, there were 140 *Callicarpa* trees. Their fruit was available year round, which probably accounts for the fact that 16 percent all feeding observations occurred in these trees. The fruits are round and small, about half a centimeter in diameter, and the animals preferred to eat them ripe, after they turned from green to red. There were typically about a dozen ripe clusters of these fruits on a tree, and they were harvested in about five minutes by one individual. Their percentage of crude protein was rather high, at 11.35. The diet of the study troop was estimated at 87 percent fruit, and it became apparent that one explanation for the animals' ranging in the streamside areas was this relatively continually available and more predictable fruiting compared to the fruits of the primary rain forest. Surveys of fallen fruit on 1,800 meters of trails, twice a month over a period of a year and a half, showed that fruit was significantly more frequent and continually available in streamside areas than in nonstreamside areas (Mann-Whitney-Wilcoxon Tests p.<.05) (Wheatley 1978b). The larger trees, characteristic of primary rain forest, very infrequently contained fruit; but when they did, the troop revisited these trees day after day for weeks until the crop was exhausted. One fruiting *Koordersiodendron* tree was visited continually for almost two months.

Another aspect of this macaque's diet, which facilitates its adaptation to secondary habitats, is its dietary diversity. The Shannon-Wiener diversity measure of the forty-five different kinds of fruits for which local names were known was 3.19. Adding in the other fruits for which local names were not known would raise this figure to 3.42 (Wheatley, 1980b). P. Lucas and R. Corlett (1991) and Carey P. Yeager (1996) also report this diversity of dietary fruit for the species. In sum, the important factors according for the monkeys' success in riverine secondary forests, at least in the Kutai, is their dietary diversity, relatively abundant and continually fruiting riverine trees and their large home range of

1.25 square kilometers for this relatively large troop of thirty animals.

The preference of this troop of macaques for secondary forest, especially in a riverine area, is not unusual for the species. Quite the contrary, most surveys report that the preferred habitat of this macaque is the same as the one that I documented (Aldrich-Blake, 1980; Crockett & Wilson, 1980; Marsh & Wilson, 1981; Medway, 1970; Rijksen, 1978; Rodman, 1978; Southwick & Cadigan, 1972; Sussman & Tattersall, 1986; Wilson & Wilson, 1975; 1977). Troop sizes, for example, are generally larger in the more disturbed forests such as riparian, mangroves, and secondary forests than in the primary forests (Crockett & Wilson, 1980; Wilson & Wilson, 1975). Andrew and Bettina Johns (1995) call *M. fascicularis* "a specialist of high-productivity edge and secondary habitats." In a restudy of a logged forest in peninsular Malaysia, they found a rapid recovery of long-tailed macaques after logging and an extension of their ranges into neighboring logged forests.

The effect of humans on the behavior of the animals in this "undisturbed" troop in an "undisturbed" habitat was apparent in another way. For many months after my arrival, the animals would hide, freeze motionless, or silently flee whenever they detected my presence. The animals also left their sleeping tree earlier and earlier in the morning, as if trying to leave before my arrival. Walking in the forest before sunrise was always interesting. Tarsiers returning to their nests would freeze in the beam of my flashlight long enough for me to touch them. Barking deer lived up to their name when, about a meter away, they sprang up out of concealment and let loose with their heart-stopping bark. The sun bears also barked when I surprised them. Eventually, I showed up over an hour before sunrise, while it was still too dark for most of them to leave the tree. Some of them would coo, unable to find their mothers, or would jump into the branches of other trees in the dark. Such behavior is very different from that of other "animals." Their reaction to the sun bear, for example, was to mob the bear until it left. It was only after a year of observations on this troop of monkeys that the animals mobbed me, and this occurred when I climbed into a tree to make better observations of them. Whether this was a statement of their perception of me as an animal now, in contrast to a year ago, I don't know. One local villager remarked in jest (I think) that with my red beard, I looked just like an orangutan. I doubt that any other humans climbed trees near them.

The Antiquity of Primate Commensalism in Southeast Asia

The antiquity of this association between humans and this species of monkey is an interesting one. Monkeys and other mammals have long shared the various species of fruits and tubers indigenous to the area. Bananas, coconuts, rambutan, durian, breadfruit, as well as taro and yams were available to both monkeys and humans. The regrowth of secondary forest areas after the abandonment of ladang, or "dry" farms, which are nonirrigated, surely provides new habitat for monkeys. Their success and dispersal following the footsteps of humans would thus increase. This commensalism would therefore be expected to be at least as old as ladang and other types of agriculture. D. Walker (1980) and Peter Bellwood (1980; 1993), for example, state that agriculture and horticulture go back 4,000 to 5,000 years ago. Prior to the introduction of rice, traditional ladang of Bornean dayaks focused on taro and sago (Christensen & Mertz, 1993). Horticulture utilizes vegetative replanting, and the cultivation of crops such as yams, taro, sugarcane, coconut, sago, and various species of bananas have long been part of Southeast Asian horticultural systems. The monkeys in Bali raid the farmers' crops, eating yams and taro, and the latter crop is also eaten on the Island of Ngeaur in Palau (Wheatley, 1988). Taro (*Colocasia esculenta*) was probably grown in New Guinea at least by 7000 B.C., and since the plant is not native there, but to Southeast Asia (White, 1984), it must have been used much earlier. Some of the other staple crops that monkeys eat, such as rice, cassava, and sweet potato, were introduced later.

Lord Medway (1972) suggests that humans brought some of the commensal animals with them, namely the rat, the house mouse, wild pigs, possibly civets, and sambar deer. Did humans ever bring monkeys? A report by Ian C. Glover (1971) says that humans introduced *Macaca fascicularis* to Timor some 4,000 to 5,000 years ago, along with deer, civet cats, and cuscus. Interestingly, this is about the same time as the introduction of agriculture and the introduction of such domesticated animals as dogs, goats, and buffalo. It is not far-fetched to suggest that humans brought monkeys as pets or food that long ago. One of the oldest documented cases of animal introduction in the world is the gray cuscus, an arboreal marsupial on the islands of Bismarck at about 18,000 to 20,000 years ago (Kirch, 1997). This species is still popular as a pet and for food. There are many examples of the introduction of *M. fascicularis* onto various islands, probably as pets which later escaped. Among the more famous examples are the introduction and success of the species on Ngeaur in Oceania and on the island of Mauritius in the Indian Ocean. It is not unreasonable to assume that the species was probably introduced to many of the islands of Indonesia in the past, as several researchers such as A. Hoogerwerf (1970) and H. Sody (1949) have stated. Germans introduced *M. fascicularis* onto the island of Ngeaur around 1910, and the Dutch may have been responsible for the introduction of *M. fascicularis* onto the island of Mauritius at least by 1606 (Sussman & Tattersall, 1981). There are interesting accounts by François Valentijn (1994 [1724–26]), who reports that three ships from Holland anchored at Bali in 1597. The fleet was under the command of Cornelius de Houtman, and the crews went ashore many miles inland. They also loaded pigs, ducks, fowl, eggs, and fruits onboard ship. Several Dutchmen went ashore and never returned to the ships. The ships' names were the *Mauritius*, the *Hollandia*, and the *Duifken*. Monkeys are not mentioned in these reports, but there is the possibility that Balinese macaques may have been brought rather than or in addition to Javanese macaques, as Robert Sussman and Ian Tattersall (1981) have reported. Both subspecies are very similar. Another possibility is that the Portuguese or Arab traders introduced the macaques at an even earlier date (Sussman & Tattersall, 1981).

Interestingly, some of the earliest evidence of both humans and *M. fascicularis* in Indonesia and Southeast Asia occur together in the same deposits in Java (Hooijer, 1952; Medway, 1972). Many other mammalian fauna are here, including primates such as langur, gibbon, siamang, and orangutan. Dates of 1.66 and 1.81 million years ago at the villages of Sangiran and Modjokerto, respectively, have recently been published for these deposits (Swisher, Curtis, Jacob, Getty, Suprijo, & Wisiasmoro, 1994). Pollen analysis of Sangiran marine estuarine sediments shows mangroves, nipah palms, and pandanus trees in the Pliocene and extending into the Pucangan period, when they were gradually replaced by rain forest and more open country (Bellwood, 1980; 1997). The entire faunal assemblage, known as the Sino-Malayan fauna, implies an ancient floral mosaic of forest, scrub and grassland, namely, a sort of heavily wooded savanna.

Even earlier mammalian faunas called Siva-Malayan exist in Java at Tji Djulang and Kali Glagah. These early lower Pleistocene remains include such fauna as the southern mammoth *Archidiskodon*, *Leptobos*, and *Equus* (Medway, 1972), but no primates. These and other large herbivores are characteristic of riverine forest-edge habitats. It has been suggested that macaques dispersed to Sulawesi during this time (Bellwood, 1997).

The entry of mammals into the Indonesian archipelago was also dependent on the dynamic changes of water levels and volcanic activity during the Pleistocene. The archipelago is home to the Sunda shelf, an area of almost 2 million square kilometers, presently the shallow Java and South China seas and other areas such as between Java and Bali that are now submerged less than 100 meters (Ollier, 1980; Walker, 1980). A drop in sea level of only about fifty meters would be enough, for example, to unite the present-day islands of Borneo, Java, and Sumatra with peninsular Malaysia. The Sunda shelf is the most extensive shelf in the world. When the shelf was exposed as land, perhaps as many as twenty different times throughout the last few million years until about 6,000 years ago, it formed an extensive continent known as Sundaland. The many cycles of sea-level changes forced an alternating habitat between continent and archipelago that would favor colonizing species, including not only mammals such as *M. fascicularis*, but also plants, for example, pioneering mangroves, *Rhizophora* and *Avicennia*. Such species have no problem adjusting to these rapid changes in sea-level (Ashton & Ashton, 1972). The Ashtons estimate the rate of coastline change at about ten meters per year over the last 5,000 years for the Limbang and Baram valleys based on present-day sedimentation of the river mouths. *M. fascicularis* has consequently adapted well to these naturally disturbed habitats over the last 2 million years. At the present time, humans produce these disturbed habitats, and *M. fascicularis* has rapidly accommodated to them.

With such ancient ecological associations between nonhuman primates such as *M. fascicularis* and humans, what other types of associations were there? In a recent article, Leslie Sponsel (1997), for example, says, "human predation on nonhuman primates has been grossly neglected, as if *Homo sapiens* were not a natural part of the faunal community of forest ecosystems." Despite the major threat that hunting can pose to primates, such as local extinction in some areas (Mittermier, 1987), there is a reluctance to consider humans as "natural" predators. Monkeys are an important source of protein for some indigenous human populations in Amazonia, and they rank in the top three categories of numbers and weight of mammalian prey (Sponsel, 1997). Human predation has probably had an important effect on populations of monkeys for 6,000 to 8,000 years in the Amazon. In addition, swidden agriculture may also be that old in the Amazon basin; it thereby affects the dispersal and success of some of the monkey species, such as howlers and squirrel monkeys, which are attracted to secondary forests (Warren Kinzey, 1997). Some indigenous peoples such as the Kayapo will even plant trees in order to attract game into the fallowed gardens.

There is no evidence of human hunting in the Jetis and Trinil deposits. However, monkeys and other animals appear to have been eaten at Niah Cave in Malaysian Borneo by 40,000 years B.C. The cave deposits at Niah Cave are radiocarbon-dated back to 40,000 years B.C. (Hooijer, 1962; Medway, 1958; 1964). The most common bones of a mammal that was eaten then appear to be those of *M. fascicularis*, which are found at some of the lowest excavated levels in the cave. There are no human weapons associated with these remains, but Medway (1958) indicates that they "are easily trapped on the ground." Hooijer (1962) further notes that the macaque cave deposits were clearly brought in for food by humans. Present-day Penan in Gunong Mulu National Park of Sarawak (Malaysian Borneo) hunt and eat *M. fascicularis* as well as other primates (Labang & Medway, 1979). Other areas in Southeast Asia also show early evidence of monkeys used for food. Excavations in Sulawesi, for example, show that monkeys were eaten around 8,000 to 10,000 before the present era (Glover, 1984). Monkeys were hunted in China about 5,000 B.C. based on deposits found at Hemudu, which also include large quantities of rice (Bellwood, 1997). Cave deposits in peninsular Malaysia such as Gua Cha in Kelantan, also have monkey bones that date between 10,000 and 3,000 years ago (Adi, 1985). The young of these and other animals, especially pigs, may have been a favorite food species not only at this site, but in many Hoabinian sites in general (Bellwood, 1997).

Ngeaur Island, Republic of Palau

One of the most destructive effects humans can have on nonhuman primates is hunting. Although subsistence hunting of primates with bow and arrow and blowpipe can seriously affect primate populations, the use of firearms is potentially devastating. It is understandable that primatologists are reluctant to conduct research under fire, but it is important to know how monkeys might try to adapt under such circumstances. This reluctance can seriously affect the validity of a number of hypotheses, such as sociality, sex roles and other responses to predation.

Despite the difficult observation conditions on Ngeaur (Poirier & Smith, 1974) and its past history of human hunting (Farslow, 1987), I decided to accept the invitation of Rebecca Stephenson and Hiro Kurashina at the University of Guam. After a month of training nine students in the summer of 1994, we went to the island of Ngeaur. Rather than avoiding the study of human predation on monkeys as most primatologists do, we deliberately sought to study it.

Ngeaur is a small (830-hectare) coral atoll. Despite our limited time, only ten days on the island, the twelve of us conducted over 700 hours of survey work and observation. We tested a number of hypotheses on the effects of human hunting on macaque social behavior. For example, it is commonly believed that individuals of larger groups are safer, that such groups should be multimale and that they can detect predators from a longer distance than smaller groups. In addition, males are supposedly more vigilant than females, quicker to communicate the presence of predators, and more likely to approach and mob predators. We thought that humans hunting with firearms might pose different problems to the macaques than other so-called "natural" predators. We also took a census of the macaque population by conducting line transect surveys.

The macaques of Ngeaur are the most easterly distributed wild population of any nonhuman primate species. They are also the only wild monkeys in Micronesia and, indeed, of all of Oceania. We corroborated the previous report of Frank Poirier and Euclid Smith (1974) that the animals were initially brought to the island during the German occupation, probably between 1909 and 1914. Kurashina's interview in Japanese of Ucherbelau Masao Gulibert, the chief of the top-ranking matriclan, however, revealed that five animals, rather than the two originally reported, were brought by ship from Indonesia. This is more in line with the genetic tests by Kiyoaki Matsubayashi, Shunji Gotoh, Yoshi Kawamoto, Ken Nozawa, and Juri Suzuki (1989) that showed too much genetic diversity for the current animal population to be descendants of just one pair. In addition, our interviews also revealed that the Germans had these monkeys in their hospital, apparently for drug or disease testing, rather than in their phosphate mines as the proverbial "canary." The mining activities were all on the surface, especially during the first few decades of activity, so use of the monkeys as canaries in the mines is very doubtful.

The local people consider the macaques of Ngeaur a major agricultural pest. Crops such as bananas, oranges, and taro are eaten and damaged. At the time of Poirier and Smith's study in 1974, the monkeys were seldom hunted, and troop sizes averaged forty to fifty animals. Legislation passed in 1975, however, called for the eradication of monkeys, and rewards were offered for their tails. Even after the 1979 Constitution of Palau outlawed the use of firearms, Daniel Farslow (1987) states that 10 percent of the population (ninety-four animals) were shot in 1980. He did not observe the trapping of animals that goes on today (Farslow, pers. comm.). Farslow (1987) estimated a total population of between 825 and 900 animals living in thirteen troops, averaging fifty-two animals per troop.

During our stay on Ngeaur, the Koror police shot three monkeys, using M15 military assault weapons and shotguns. Our population estimates based on surveying over 62,000 meters of roads and trails, plotting troop locations on maps and counting individuals yield a figure between approximately 350 to 400 animals. This figure is half as much as fourteen years ago. We also found over twice as many troops, with significantly fewer individuals in each troop. The same areas that averaged fifty-two animals per troop fourteen years ago now averaged eleven individuals per troop. We found twenty-seven troops, compared to Farslow's (1987) thirteen troops. It seems reasonable to conclude that hunting and trapping are probably responsible for the population decline. We did not hear or see any evidence of the local people eating these monkeys. Why bother eating monkeys when fresh fish such as yellowfin tuna and many different kinds of crabs seem readily available? Rather than worry about fixing our own food, we went snorkeling next to our guesthouse and relied on the superb cuisine and hospitality of the islanders.

Our data from Ngeaur challenge some of the proposed hypotheses accounting for primate defense against predation. Obviously, human hunting yields much higher rates of mortality on *M. fascicularis* than other documented predators, such as the gavial, the Philippine monkey-eating eagle, and the python (Galdikas & Yeager, 1984; Rabor, 1968; Van Schaik, Van Noordwijk, de Boer, & den Tonkelaar, 1983). On these occasions, only single animals were killed. Our data do not support the hypothesis that larger groups avoid predators better than smaller groups. The troops we surveyed had significantly fewer individuals than those surveyed by Farslow (1987) and by Poirier and Smith (1974). Furthermore, we found no significant correlation between the total number of animals in the group and detection distance. Larger groups do not appear to detect human hunters any better than smaller groups. Our interviews of hunters on Ngeaur revealed that they were more successful when they hunted the larger troops. Dogs were sometimes used in hunting monkeys. The dogs and hunters would chase and surround the monkeys in trees and then shoot them. Hunters preferred to shoot the mothers of infants in order to sell the surviving babies for U.S.$100.

Not all areas on Ngeaur were hunted with equal intensity. The most frequently hunted area was next to the gardens, especially in the swamps, where the taro grew. We decided to focus our observations on a troop near the Catholic Church, which was in the most frequently hunted area, and to compare these results to our results from observing another troop that was said to be free from hunting. The latter troop ranged near an old and partly destroyed Japanese-built lighthouse. Several hunters told us that they did not hunt in this area because there were no gardens, and it was too rocky. Both troops were multimale and approximately equal in size (fifteen to sixteen individuals) and in age-sex composition. We found a number of interesting differences in social behavior between the two troops, which probably relates to the differences in hunting. The most obvious difference was that the troop more exposed to hunting was significantly quieter and more evasive than the other troop. The church troop had significantly fewer alarm calls, fewer calling individuals, fewer animals visually detected by the observer, fewer exposed individuals, and fewer animals that threatened or approached the observer. Some of their alarms were barely audible. Most of our first contacts with monkeys were by tracking their alarm calls, and it seems reasonable to assume that calling animals attract the attention of hunters and dogs such that they would then be tracked and shot. We did one callback experiment in which we tried to simulate a hunt at the lighthouse with a playback of a barking dog. The animals fled and became more silent when hearing this sound.

Human hunting with firearms thus has some different responses by monkeys compared to other nonhuman predators. The Ngeaur macaques do not quickly communicate the presence of human predators in contrast to the same species in Sumatra when it is exposed to nonhuman predators (Van Schaik, Van Noordwijk, Bambang Warsono, & Sutriono, 1983). We also did not see a significant difference in vigilance between the sexes as the above researchers found. Vigilance by females appears to be more important when animals are subjected to human hunting perhaps because the location vis-à-vis other animals makes less difference to being shot than simple exposure. One might, for example, expect nonhuman predators to hunt more peripheral animals. We also found no confirmation of the hypothesis proposed by the above researchers that Asian macaques outside the Philippines should live in troops with only one adult male because of the absence of large raptors. Both the lighthouse and the church troops that we observed were multimale.

Our experience on Ngeaur has again shown us how important humans can be on the behavior and social organization of this species of monkey. First, the successful colonization of macaques on Ngeaur was the direct result of humans, Germans, who brought the animals to the island some eighty years ago. Paradoxically, as despised as the Germans are by the local people for this event, the local people are also responsible for the selling and/or transmitting of these animals as pets to some of the other islands of Palau. The pets have escaped on these islands, but they have not apparently colonized yet. Second, by feeding caged, captive animals on Ngeaur such novel fruits as papaya, oranges, and bananas, the local people may be making the crop-raiding problem worse because the captives sometimes escape and have learned to feed off local crops. Third, human hunting of these animals with firearms produces different effects than other predators do on the social behavior of these macaques. The population size has decreased by half, down to about 400 animals, and the numbers of troops appear to have doubled while their sizes are significantly smaller than reported ten years ago. One troop subjected to more human hunting is quieter and more evasive than another troop subjected to less predation by humans.

The effect of human hunting on primates needs to be included in our attempts to construct hypotheses on the effects of predation because human hunting has and will continue to have marked effects on nonhuman primates. Last, in contrast to the Balinese, who have a long tradition of primate commensalism, the local people of Ngeaur are waging a battle to destroy the macaques, even though the animals raid crops on both islands. Almost every

Ngeaurese that we talked to hated the monkeys and wanted them destroyed. This difference shows the importance of community support for the effective conservation of monkeys. Despite the ban on firearms, monkeys are still shot on Ngeaur, and despite the U.S.$100 received from the selling of monkeys off the island, the local people perceive the monkeys to be detrimental to their economic well-being.

The Monkey Forest in Padangtegal, Ubud, Bali

The association of both monkeys and humans in riverine areas of Borneo, Bali, and elsewhere accounts for the monkeys' reputation as pests. They raid the riverine gardens of Borneo, eating rice, beans, corn, cucumbers, and other crops. Farmers built huts next to their gardens to spend the nights, and they would guard their ripening crops from monkeys and other animals such as pigs (pers. obs.). The Dutch labeled these monkeys as "harmful vermin" throughout Indonesia, according to the Animal Protection Ordinance and Regulations of 1931, Nos. 134 and 266 (Hoogerwerf, 1970). Another measure of the success of this species is its adaptability to habitat change, for example, to human logging, as Johns and Johns (1995) have shown. In Bali, and elsewhere such as Thailand, this species of macaque has been living and associating with humans for centuries at Buddhist and Hindu temples (Angst, 1975; Eudey, 1994; Wheatley, 1988). By studying such populations in a Balinese temple forest environment we can expand our knowledge of this species' adaptability.

The Balinese populations of *M. fascicularis* belong to a unique subspecies, according to H. Sody (1949), which he called *M. irus submordax*. This subspecies is not, however, recognized by Jack Fooden (1995), who places them in *M. f. fascicularis*. The Balinese *fascicularis* have a tendency for insular dwarfism. They are, for example, smaller in head and body length and significantly smaller in skull length than the animals in the so-called core areas, namely Java, Borneo, Sumatra, and the Malay and Indochina Peninsula (Fooden, 1995). The old species' names for *fascicularis*, such as *cynomologus* and *irus*, have been dropped according to the rules of zoological nomenclature (Blanford, 1887; 1942; Fooden, 1964; Napier & Groves, 1983) in favor of *fascicularis*. Sody retained W.C.O. Hill's (1974) subspecific name of *mordax* for the Javanese population of macaques. Recent electrophoretic examinations of blood proteins have confirmed this subspecific differentiation (Kawamoto & Ischak, 1981; Kawamoto, Nozawa, & Ischak, 1981; Kawamoto, Ischak, & Supriatna, 1984; Kawamoto & Suryobroto, 1985).

Some twenty kilometers north of Denpasar, the provincial capital of Bali, is the Monkey Forest of Padangtegal. It is located just south of Ubud where, appropriately enough, Anoman Street intersects with Monkey Forest Street. The monkeys at the Monkey Forest in Ubud, however, have been living next to people for hundreds of years. According to my informants the Pura Dalem, the temple for the dead, is at least 400 years old; and monkeys have always been there. I was interested in learning what types of interactions existed between

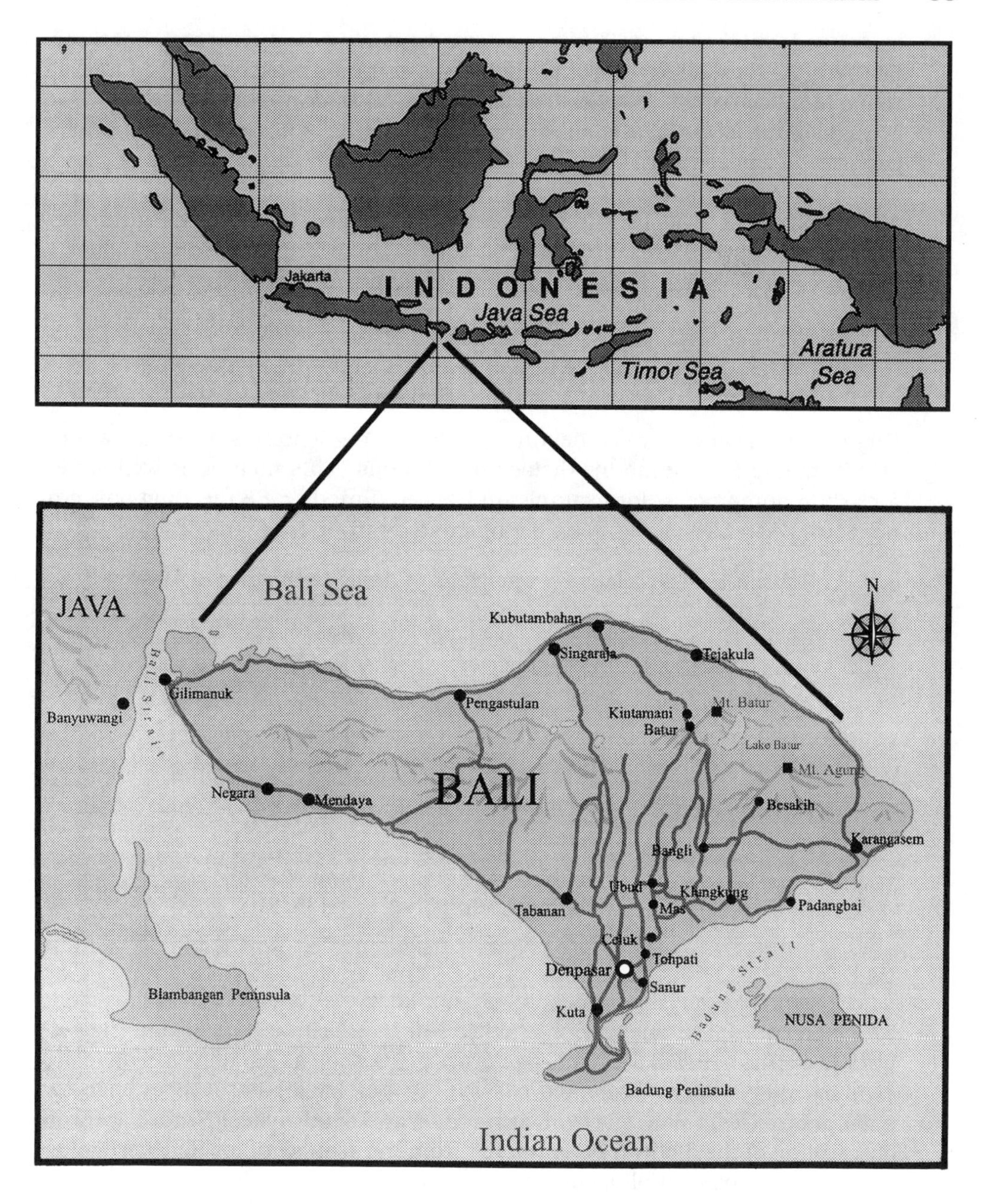

these monkeys and the Balinese. How dependent were they on human sources for their diet?

My first summer's observations on the three troops in this forest sanctuary began on June 12, 1986, and ended on August 16, 1986. Approximately 341 hours of observation were made in fifty-seven days. The troop I chose to study most intensely was usually near the road called Anoman Street, at the eastern edge of the forest. This troop turned out to be the most dominant troop. In 1986, the local villagers from Padangtegal commonly used this road to walk to the Monkey Forest. Another troop, the second-ranking, ranged on the western side of the forest. The smallest troop spent much of its time in the southern part of the eastern gorge.

The remaining research was conducted during the summers of 1990, 1991, and 1992. The 1990 session began on August 1 and lasted until September 20. The 1991 research session started June 2 and ended on August 2, and the 1992 session lasted from June 22 until August 6.

Diet of the Monkeys in the Monkey Forest

Dietary observations were begun in 1986, using Thomas T. Struhsaker's (1969) method of estimating dietary proportions. This method is well suited for arboreal monkeys, for example, in Borneo. I made 5,883 feeding observations, of which 328 were not on the main study troop (see table 2.1).

Table 2.1. **The relative proportions of items in the diet in 1986**

Item	Proportion
Peanuts	23.2%
Sweet potatoes	19.4%
Fruits	17.9%
Insects	12.5%
Grass	10.9%
Offerings	8.7%
Leaves	4.6%
Flowers	2.0%
Pith, roots, sap	.8%

The most noticeable aspect of the monkeys' diet is that humans provided 57.7 percent of it in one way or another. Tourists primarily provided the peanuts as well as some of the fruits and other items such as bananas, bread, cake, candy, cookies, corn, fermented cassava, grapes, jackfruit, oranges, papaya, pineapple leaves, salak, starfruit, sugarcane, and sandwiches. A more accurate description of the diet can be estimated from the scan samples that were taken on all three troops simultaneously by three observers every half-hour at the end of each summer's research in 1990, 1991, and 1992. These scan samples recorded the activities of each visible monkey, determined within fifteen seconds of viewing the animal. We would try to note as many animals as possible within a five-minute time period. In 1990, for example, tourist food provided about 44 percent of their diet and offerings about 4 percent. These estimates are for all three troops. The third-ranked troop (Troop Three), for example, rarely ate tourist food because it generally ranged beyond the roads where tourists frequented. For the most dominant troop (Troop One), the 1990 scans show that about 56 percent of their diet came from human sources. In 1991, the scans show that tourist foods provided 69 percent of the diet to Troop One, 73 percent to Troop Two, and 39 percent to Troop Three. In 1992, Troop One's dietary percentage of tourist food was 59 percent, while Troop Two's was 80.5 percent and Troop Three's was 79 percent.

Local people also provided food, especially the sweet potatoes, which were generally provisioned by the guards at the feeding station (see Plate 2.3). Such provisioning was said to have started in about 1976. The rationale for provisioning was to allow tourists to see the animals as well as to keep the monkeys from raiding the farmers' fields. At that time, handicraft vendors also provided peanuts to the monkeys to facilitate peanut sales to tourists.

Offerings, 8.7 percent of the monkey diet, are food items left at the graveyard by local people for the various gods, demons, and spirits, especially the latter. Balinese typically cremate their dead. Commoners (*sudra*), however, sometimes temporarily interred their dead until cremation could be provided. The corpse would be wrapped in a mat and buried with a bamboo tube placed under the head to let out the spirit. The graveyard was on the south side of the cremation temple. Sometimes the family's (especially the children's) obligation to cremate the remains of a parent are not fulfilled at all, and the bones of these uncremated bodies can be seen eroding from the graveyard. The most common offering placed next to and on the grave was cooked rice placed on banana leaves. Other offerings included bananas, bean sprouts, corn, cucumbers, krupuk, noodles, oranges, pongi nuts, salak, sate, and various flowers. The monkeys sometimes competed with dogs for these offerings and would drive the dogs away.

One aspect of the monkeys' diet is crop raiding. The animals raided the surrounding fields on eight different days in 1986. On six of those days, they raided the padi fields seven times. They also pulled up and ate rice stalks, sweet potatoes, and numerous other crops and helped themselves to the numerous fruits growing in the gardens (see Plate 2.4). The farmers generally detected the monkeys after an average of 32 minutes and chased them out. The farmers threw rocks and ran after them with sickles. Troop One's second-ranked adult male had a very bent tail reportedly caused by a sickle. In 1992 I saw a fresh, long, and deep gash across the thigh of the dominant male of the third-ranked troop. When I spotted him, adult females were at the edge of a gorge near a field, grooming him. Some of the females threatened me, and one ran up to me. I was unable to back away because of a deep gorge on all sides, and I simply let her bite me. It was only one bite, and it did not draw blood through my pants. The bite struck me as redirected aggression from a farmer onto me in a symbolic sense. I feel confident that the result would have been much more severe if I had resisted, thereby risking a mass attack by coalitions of animals, especially the females. In 1991, I saw this same male repeatedly threaten and circle two farmers in the paddy field, one of whom had a hoe and the other a sickle. No one was injured in that encounter. Some farmers used long poles to beat the trees where monkeys rested at the edge of the sanctuary.

Plate 2.3. *A wounded alpha male, Two Fingers of Troop One, holds a sweet potato near the front entrance to the Monkey Forest*

➤

Initially I followed the monkeys on their crop raiding to make detailed feeding observations. Later on, however, I watched from the forest. When the farmers discovered me hiding in the forest, after they had chased the monkeys, they were extremely embarrassed. Years later they admitted that they were actually afraid of me and were surprised to find that I was sympathetic to their problems with the monkeys. These farmers were often newcomers to the village, working land that few others would want, being so close to the monkey forest. Further details on crop raiding can be found in Wheatley (1988; 1989).

Monkeys can eat great quantities of padi leaf blades. One female, for example, pulled up 40 padi stalks and ate their bases in a four-minute period. Some of the farmers seemed to be even more perturbed when the animals only stepped on the rice plants to keep their feet out of the mud. The monkeys also pulled up and ate other crops such as sweet potatoes (*Ipomoea batatas*), cassava root (*Manihot esculenta*), and even taro root (*Colocasia antiquorum*). The techniques of food preparation are interesting. All of the sweet potatoes and cassava that I observed being eaten were always peeled with the incisor and canine teeth. They were also often washed and rubbed in the hands as well (see Plate 2.5). The peeling of sweet potatoes and cassava probably relates to the high levels of certain toxins that cause prussic acid poisoning. These toxins are higher in the cortex than in the rest of the flesh. Taro root contains calcium oxalate crystals and can cause a burning sensation in the mouth and throat (Wang, 1983). It is interesting that these techniques of food preparation of potentially toxic items is an adaptation to the deliberate attempt of farmers to plant such crops around their fields in hopes of conditioning the monkeys not to raid their crops. The effects on monkeys eating cassava warrants further investigation. These effects range from goiter to neurological disorders (Ermans, Mbulamoko, Delange, & Ahluwalia, 1980). The most subordinant troop, Troop Three, appeared to have more physical problems than any other troop. Ranging on the periphery of the Monkey Forest, it was more exposed to injuries and its members may have eaten more cassava (see chapter 3).

The monkeys ate the fruits of many crops. The fruits in the Monkey Forest belong to the Pura Dalem. Most of the coconuts probably never mature because the juveniles eat the base of the young fruits. The adults prefer the mature fruits outside the forest, which grow along the edges of the fields. The harvesting and husking of coconuts is a time-consuming job, sometimes taking twenty minutes. Adults will remove the husk by biting and pulling it from the stem to the bottom about fifty times over six minutes. Sometimes the adult males will wait patiently until another animal finishes the job of husking the coconut, and then he will take it away for himself. The monkeys also drink the milk and scoop out the gelatinous meat. The older and harder coconuts given to them by tourists or local Balinese can be a challenge to open. One adult female, Black

Plate 2.4. *This adult male eats the grass blades of rice in a rice field.*

Mandible, ran over and took one away from a juvenile. She then stripped leaves from a nearby tree and rubbed the coconut. Twelve minutes later, she had finally opened it by slamming it onto the ground eight times and finally twice more on the cement. She had repeatedly pushed her infant away during this period, seemingly determined to open the coconut.

The animals also eat salak fruits and the areca nuts, or *pinang*, that is used in making betel, both of which are found in the forest. All of the fruits grown in the nearby gardens were eaten. Banana plants were pulled down and even small green bananas were eaten as well as breadfruit and papaya. In addition, tamarind, green beans, nangka, papaya, jack fruit, and coconuts were also eaten. Traditionally, then, monkeys have probably been raiding crops for as long as they have been associated with human agriculture. These themes recur in the various Balinese folktales such as "Boosard, der Affenkönig" (Boosard, the Monkey-King), translated by Jacoba Hooykaas (1963).

Other dietary items include fruits, insects, and grass. If all human food sources are totally excluded from the diet, then fruit, at 32 percent, ranks the highest in dietary proportions, with insects ranking second at 29 percent, followed by grass at 23 percent, leaves at 11 percent, and flowers at 5 percent. These data are based on a total of 2,482 feeding observations during 1986. Most of the natural fruits eaten were figs, especially the ripe beringin and kresik. The animals did not eat the greener, unripe figs. Even with such good observation conditions, the species' identification of insects was difficult. Many of them were caterpillars, grasshoppers, and larvae of different kinds. A future study on the role of monkeys in controlling insects is warranted, especially on those insects responsible for crop damage and disease. The stalks, pith, blades, and rhizomes of grasses were also eaten. The animals not only ate the

Plate 2.5. *A juvenile monkey washes this cassava root again before taking another bite. Note the leaves of this plant on the ground in the lower right corner and the rice field in the background.*

bases of rice stalks but also the blades of other grasses such as *Cynodon* or *Zoisia* in the rice fields.

Other dietary items were flowers, pith, sap, and snails. Snails were eaten by using the fingers to scrape them out of the shell. The flowers of cauliflorous trees were eaten as well as the insides of hibiscus flowers. The pith of trees, especially from the boh trees, was eaten, and sap was also licked. An adult female alternately bit a tree branch with her incisors and licked the sap. An adult female was also seen to use her incisors to bite the bark off a tamarind tree and lick underneath. Perhaps the most unusual dietary behavior is temple licking (see Plate 2.6). Bouts of temple licking averaged almost 13 minutes, and females had significantly longer bouts than males. Females averaged 8.1 minutes, whereas males averaged 3.5 minutes ($N = 33$). Most likely, essential elements, especially salt, which percolates down through the temple wall after rainstorms, were being obtained during temple licking. Osamu Takenaka (1986) reported marginal deficiencies in the blood plasma of monkeys in a nearby area. Extra sodium is necessary for adult females during pregnancy and lactation.

Plate 2.6. *Three female monkeys lick the walls of the Cremation Temple in the Monkey Forest.*

There is no doubt that the Monkey Forest at Padangtegal has a high level of human disturbance. On the four-point scale proposed by N. Bishop, S. Hrdy, J. Moore, and J. Teas (1981), the degree of human disturbance at the Monkey Forest averages 3.25. The size of the forest sanctuary is about four hectares. Including the additional ravines and rice fields that the animals were seen in, their total range is about ten hectares. With a population of 134 animals in 1992, the density of individuals is very high: 1,340 per square kilometer. The animals were provisioned on a daily basis by local villagers and/or tourists while we were there. The proportion of dietary items from human sources is high.

Cultural Behaviors of the Monkeys in the Monkey Forest

An early Western attitude toward monkeys and apes was that their behavior was stupid and even evil when they imitated human behavior. By focusing exclusively on the differences between humans and apes, which helps to drive apart and to separate us from them, we continue to erect artificial boundaries that prevent us from evaluating our commonalities with nonhuman primates, especially when it comes to issues of culture and intelligence. Rather than dismiss this "intelligent" behavior as stupid, we need to examine it in more detail. If our anatomy, morphology, and genetics are very close to those of apes and other primates, why not also our behavior? The behavior of the monkeys in the Monkey Forest at Padangtegal is therefore useful to study in this regard.

The definitions of culture in the many introductory cultural anthropology books are interesting, not only in their varying definitions but also in their commonality. In other words, culture is whatever humans do. These definitions do not help us much when we want to know if apes, monkeys, or even fossil hominids have culture. Some researchers, such as Michael Huffman (1984), A. L. Kroeber (1928), and William McGrew & Carolyn Tutin (1978), have attempted operational definitions of culture based on certain components. These components include innovation, dissemination, standardization, durability, diffusion, tradition, nonsubsistence, and natural adaptiveness. All of these components are seen in another species of macaque, *M. fuscata*, the Japanese monkey. Japanese primatologists have called the discovery of cultural behavior in these monkeys one of the "brightest achievements accredited to Japanese researchers in primatology" (Kawai & Ohsawa, 1983). This refers to the washing of sweet potatoes in the Koshima Troop, which is often cited as one of the standard examples of nonhuman culture. The washing of sweet potatoes is "proto-tool-use" according to Sue Parker and Kathleen Gibson (1977). Anthropologists also consider the rubbing or "washing" of food items in leaves as tool use, another component of culture (Beck, 1980; Chiang, 1967).

These cultural behaviors are also seen in the monkeys of Padangtegal (Wheatley, 1988). The washing and eating of sweet potatoes and cassava roots was seen during crop raiding when water was available in irrigation canals or puddles. Animals were seen to dig these foods up, to wash and to peel them with their teeth and eat them. It should be pointed out that in most cases, the sweet potatoes appeared to be cleaned, but that this was not always the case. Sometimes an apparently clean object was dipped in a muddy puddle that appeared to make it dirtier. Even adult animals did this, so their learning, for example, of our concept of cleanliness, seems incomplete. Another example of this occurred when an adult male ate a bag of peanuts after first washing the plastic bag full of peanuts in a mud puddle. Seventeen cases of the rubbing with leaves ripped from trees of various objects such as seeds, snail shells, coconut shells, worms, or hairy caterpillars were observed. In some cases, the leaf rubbing appears to be a cleaning behavior, and in other cases it may function to immobilize the object, such as a worm or caterpillar, prior to ingestion. Some of the rubbed objects were not edible, however, and in other cases, the rubbing precedes the animal's attempt to bite the object between the leaves, a sandwich in effect, which appeared to allow the animal to get a better grip on the object with his teeth. Hard objects such as snail shells or coconuts either had nothing in them or were not opened. During our research in later years, other examples of using leaves were also observed, such as the grooming of individuals, especially their eyes, with a leaf (see Plate 2.7). Other animals would watch individuals peeling, washing and leaf rubbing. They appeared to copy such behaviors and to incorporate the methods of leaf rubbing in play behavior. They were also the desired objects of competition by troop members. By focusing on the seemingly meaninglessness of these behaviors, however, one would ignore their potential adaptive value and their role (as tools or extractive foraging) in the evolution of intelligence and culture. Extractive foraging refers to removing food from embedded or encased objects such as a root, seed, or coconut (Gibson, 1986). Recall the observations on the adult female named Black Mandible, who cracked the coconut open by pounding it on the cement. Such behaviors appear to be learned by watching tourists and local Balinese who pound coconuts open in a similar manner and then offer them to the monkeys.

Plate 2.7. *These monkeys use leaves in many interesting ways. This animal uses a leaf to groom around the eye of another.*

Eating and foraging on such foods is adaptive in that previously unavailable food, after preparation, becomes available. Natural selection would favor such individuals who utilized tools and other intelligent behaviors.

Chapter 3

Social Behavior of Temple Monkeys at Padangtegal

Introduction

Many primatologists disdain to work in conditions where humans seem to be part of the natural landscape. Some researchers even state that their hypotheses cannot be tested on monkeys living in such "disturbed" habitats. Such statements raise all kinds of questions. What is a disturbed habitat? Are there degrees of disturbances, with some being objectionable and some allowable? Are some of these hypotheses especially vulnerable in disturbed areas? If so, what does that say about the hypotheses? Humans, of course, dominate many ecosystems directly, and there is no such thing as a habitat undisturbed by humans (Vitousek, Mooney, Lubchenco, & Mellilo, 1997). It would seem to me that if a species such as *fascicularis* thrives in disturbed areas such as temple, riverine, and secondary forests, then we should try to account for its success in those areas. There are some environments such as captive or laboratory environments, however, where so-called abnormal behaviors can be found. Such behaviors include stereotyped locomotion, somersaulting or pacing, and self-directed activities such as self-aggression, but these behaviors are generally the result of early isolation and separation from one's mother. No serious primatologist, however, has suggested that the monkeys in Balinese monkey forests exhibit such behaviors. How do the behavior and social organization of this macaque vary in its adaptation to human surroundings? Just as the results of studies of this and other species of nonhuman primates in captivity and the laboratory are important in understanding social behavior, so are the results of studies in other habitats, even so-called disturbed ones.

We can also explore aspects of the species' behavior that is very difficult to explore otherwise. Just as the studies in laboratories can give us insight into such things as personalities or allow a more experimental approach, we can take advantage of some of the features available in the Monkey Forest. The first major advantage relates to habituation. In my study of wild *M. fascicularis* in Borneo, it was only after six months of steady observation that the animals were well enough habituated for me to view them with 10 x 40 power binoculars some 100 meters away without their taking evasive action such as fleeing. It was only after a year of observations that I was able to get close enough to identify individuals, most of whom were adult or subadult males. Adult and subadult males were relatively fewer in number (six) than adult females (ten), and these males also were more conspicuous. The quantity of data under such difficult observation conditions is seriously compromised. Being bitten by malaria-carrying mosquitoes, flies, and leeches during data collection also added to the difficulties of such fieldwork. The quality of data that one can obtain under such conditions is also seriously compromised. Beyond the barest outlines, the data were insufficient to obtain a more detailed understanding of behaviors such as female dominance interactions and coalition behavior. At the Monkey Forest of Padangtegal, however, observations could be made as close to the monkeys as a meter or two (see Plate 3.1). With such a close degree of habituation, the most detailed observations could be made on this species under free-ranging conditions for the first time.

The second advantage of this study site is that there are three habituated monkey troops that are easy to recognize. With the aid of multiple observers, among them my wife, Cathleen Wheatley, students Katy Gonder, and Tripp Holman from the University of Alabama at Birmingham, and the many Earthwatch volunteers, we could easily keep track of and watch all of these troops simultaneously. Detailed observations of troop-troop behavior could therefore be made.

The macaques belong to the superfamily Cercopithecoidea, and they have a similar social system that includes such characteristics as female philopatry, cooperative and competitive matrilineal groups, strong male and female dominance hierarchies, male competition, and male emigration. By matrilineal, I mean matrifocal. Unfortunately, the word *matrilineal* is well entrenched in the primatological literature. If scientists pride themselves on the correct use of technical terms, then we should be advised to drop the term *matrilineal*. The term refers to human descent groups in which the daughters of mothers belong to a corporate group through which they claim their inheritance. Property such as houses, rice fields, jewelry, and so on is then passed from mother to

Plate 3.1. *The author takes notes on one of his subjects in the Monkey Forest of Padangtegal.*

daughter, with the mother's brother playing a crucial role. The only thing that primatologists imply when they use this term is the acquisition of the rank of the mother by her daughters. Two principles, referred to as Kawamura's principles, relate to female rank among many species of macaques and other cercopithecoids (Takahata, 1991). The first principle is that the daughter is ranked after her mother. The second principle relates to the relationship between sisters, and it is sometimes called the youngest ascendancy principle because the younger sister outranks the older sister.

The Monkey Forest is located just south of the village of Ubud between two gorges. Today, the forest sanctuary is called Mandala Wisata Wanara Wana Padangtegal. Research was done during the summer dry seasons of 1986, 1990, 1991, and 1992. The humidity is also slightly less during the dry season. Generally, the weather is more comfortable during the summer months in Bali than it is, for example, here in Alabama. There are three troops in the forest. Before discussing the behaviors of individuals in these troops, it is useful to know troop size and age-sex composition. Table 3.1 gives this information for each of the four years of our study. The troops are identified as One, Two, and Three, reflecting their original dominance order when I first began observing them in 1986.

Behavior and Dominance

As in other monkeys and macaques, this species has a rich nonvocal and vocal behavioral repertoire. Vocalizations will be dealt with at the end of this section. Many of the facial expressions and body postures and gestures have been well described for this species, for example, by Walter Angst (1975); Frans B. M. De Waal, Jan Van Hooff, and Willem Netto (1976), and Judith Shirek-Ellefson (1972). Among the more useful behaviors to record are those involved in agonistic or aggressive and submissive interactions. The agonistic behaviors are useful to note because they not only help us understand relationships between animals in the so-called dominance relationships, but also because the behaviors are used in coalitions and in obtaining certain resources. Such aggressive behaviors that we used are:

1) threat—usually involving raising the eyebrows, opening the mouth, and staring. Sometimes a threatening animal also lowered his head or did headbobs.
2) chase—pursuing a fleeing animal.
3) contact—aggression in which an animal slaps or bites another.
4) takes food from.
5) displace or supplant—an approaching animal takes over the place, often a feeding or resting site of an animal who leaves at his/her approach.

Table 3.1. **Age-Sex Composition of the Three Troops and the All-Male Garuda Boys in the Monkey Forest at Padangtegal**

		Adult		Adolescent		Juvenile			Brown Infant			Black Infant		
Troops	n	M	F	M	F	M	F	?Sex	M	F	?Sex	M	F	?Sex
A. 1986														
1	26	2	9	5	1	4	1	0	2	3[1]	0	0	1[1]	0
2	27	1	10	2	2	3	2	0	3	2	0	1	1	0
3	16	1	7	0	2	1	0	0	2	2	0	0	0	1
G. Boys	0	0		0										
B. 1990														
1	29	2	12	0	2	5	3	0	2	3	0	0	0	0
2	30	2	11	0	3	3	4	1	2[1]	3	1	0	0	1
3	22–25	3	6	0	3	—	—	9–12	0	1	0	1[1]	0	1[1]
G. Boys	13	1		12										
C. 1991														
1	32	4	12	0	3	3	3	4	1[1]	3	0	2[2]	1[1]	0
2	41	3	14	0	3	4	9	1	2	3	0	1	1	1[1]
3	28	3	9	0	2	5	4	0	2	2	0	0	0	1
G. Boys	21	1		20										
D. 1992														
1	34	3	14	0	2	4	4	0	2	5	0	1[3]	0	0
2	43	4	17	0	2	7	9	0	3	1	0	0	0	1[1]
3	33	3	11	1	1	5	6	0	4	1	0	1	0	0
G. Boys	23	1		22										

[1] Each [1] represents one death in that category observed during the study period, and the deaths are not included in the number for *N*.
M = male; F = female; ?Sex = sex is unknown; G. Boys = Garuda all-male group which was first observed in 1990. Black Infant=Newborn; Brown Infant=nursing infant.

[2] The brown female infant kidnapped from Troop Two and adopted by the alpha female in Troop One is included in the census of 1991 for Troop One. The infant's mother was probably Maggie, ranked 11 out of 14 in Troop Two.

[3] The black infant, sex unknown, in the 1992 census for Troop Two was born to Humpback and kidnapped by Kidnapper, a higher-ranking female in the troop. The baby was last seen close to death on Kidnapper and is presumed dead. The study that summer ended the day the baby was last observed on her kidnapper.

The most useful submissive expression that we used was the grin, in which the lips are retracted, exposing the teeth. One does have to be cautious, however, because some behaviors, although similar, are not exactly the same (Angst, 1975). For example, the white pout or pucker face, the grin, lip smacking, and lowering the head and "hunching" the shoulders can be used by adult males when approaching estrous females. It is interesting that males typically use these more "submissive" behaviors in their courtship of estrous females. The males generally lip smacked, hunched their shoulders, and did a pout face to indicate their interest. They usually approached the female, but sometimes upon seeing those courtship behaviors, the female would approach the male. Sometimes the males also gave an eyebrow raise.

Tables 3.2, 3.3, and 3.4 give our results of agonistic behaviors for adult females in Troops One, Two, and Three, respectively. All three tables indicate a strict linear dominance hierarchy. There are many cases that appear to confirm Kawamura's first principle, that is, that a mother's daughters rank just below her. All of these younger females form cohesive grooming groups with older females who rank above them. Some of these family groups look very similar physically. While we do not know for certain that the infants of mothers eventually grew up and entered the hierarchy according to Kawamura's principle, we strongly suspect it. In 1986, for example, Toe In of Troop One had a female infant who was eventually identified as Point Right Ear in 1990 (Table 3.2). Point Right Ear entered the hierarchy at puberty just below Toe In. In addition, we do have evidence that female infants of mothers have grown into juveniles that were seen to dominate adult females. Limpy, who was born in 1990 to Birdy Toe, the alpha female in Troop Three, and Stubby, born in 1990 to PD, the alpha female of Troop Two, both dominate adult females as juveniles in their respective troops. In 1992, Slappy, in her first year in the adult female

Table 3.2. **Outcome of Agonistic Interactions among Adult Females in Troop One**[1]

	Subordinate																	
Dominant	Al	Tb	WE	Sl	Rgb	Pb	BM	Y15	Ey	CB	ST	DE	Y8	Lf	PLE	SB	TI	PRE
Alpha 1,2,3		20	23	X	13	7	5	X	6	2	0	4	X	1	2	2	5	2
Tubby 1,2,3,4			10	9	5	2	9	2	14	3	1	8	1	7	3	5	4	2
White Eye 1,2,3,4				2	3	3	9	5	10	7	1	7	2	4	5	5	5	4
Slappy 4					X	X	5	0	2	X	1	2	3	2	3	3	0	2
Red Growth B. 12						9	X	X	3	2	X	3	X	1	2	0	0	1
Patchback #4 12							X	X	5	5	X	2	X	2	4	0	8	2
Black Mandible 34								2	5	X	6	4	4	3	12	4	3	2
Young #15 4							1		2	X	2	1	1	0	2	4	0	1
Eyelid 1,2,3,4										6	4	7	3	5	2	2	20	2
Cuttail Base 1											X	2	X	X	X	X	6	X
Scar Tail 3,4												3	4	4	2	3	1	3
Dark Eye 1,2,3,4													0	4	6	3	9	8
Young #8 4														2	2	2	2	3
Lefty 2,3,4															7	2	5	1
Point Left Ear 2,3,4																3	2	3
Spot Belly 2,3,4																	3	7
Toe In 1,2,3,4																		2
Point Right Ear 2,3,4																		

[1] The value in each cell is the frequency that the dominant won an agonistic bout with the subordinate. The numbers after each name in the first row refer to the years that the adult female was present: 1 = 1986; 2 = 1990; 3 = 1991; 4 = 1992. X = not present because of death or age (too young to be an adult).

hierarchy, ranked number three in Troop One, just under her mother, White Eye (Table 3.2). In Troop Two, Juniper entered the dominance hierarchy in 1992 just under the alpha female, PD. In addition, Daisy falls below Scarface, Humpback under Worf, and Thyme under Ada (Table 3.3). In Troop Three both Gimpy and Cleft Chin entered the hierarchy under the alpha female, Birdy Toe. All three of these females also look very similar to each other (see Table 3.4 and Plate 3.2).

The various affiliative behaviors that adult females use in approaching each other also substantiated the hierarchy. These behaviors were eyebrow raising and lip smacking, used by dominant females approaching a subordinate, and the grin when used by a subordinate in approaching a more dominant female. For example, a more dominant female appears to signal a subordinate of her "intent" to affiliate by giving an eyebrow raise or a lip smack. The subordinate animal would then typically lip smack or grin and groom or be groomed by or huddle with the more dominant animal. The following case in Troop Three is interesting because it also involves the pout face. Hot Lips, a more dominant female than Toes Up, sits on a stump and then does a pout face to Toes Up, who is ten meters away. Hot Lips then runs over and raises her eyebrows to Toes Up, who does a pout face. Hot Lips then turns her back

Plate 3.2. *A matrifocal unit in Troop Three groom and relax together. Note the strong physical resemblance of the two sisters.*

Table 3.3. **Outcome of Agonistic Interactions among Adult Females in Troop Two**[1]

	Subordinate																
Dominant	PD	JU	GY	JD	KD	AB	SF	DY	WF	HB	BV	GI	MG	AD	TH	BY	AN
PD 2,3,4		4	7	1	1	3	6	5		1	3	3	2	3	1		
Juniper 4			2	1	2			3	2	1			2	1	2	1	
Guynan 2,3,4		1		1	2	1	5	1	1	8	4			1	2	1	
Jordi 2,3,4					2	1	4	3	1	1	2	2	1	3	2		1
Kidnapper 2,3,4						3	5	6		2	2	10	1	4	1	3	1
Abby 2,3,4							1		4	2	2	2	2	2		1	1
Scarface 2,3,4								3	1	4	2	3		1		1	
Daisy 3,4									5		2		2		1	1	
Worf 2,3,4										1	1	2			1		1
Humpback 3,4												2	2	3	1		
Beverly 3,4														3			
Gimpy 2,3,4													1	1	1	2	2
Maggie 2,3,4														2	1		
Ada 2,3,4															2	1	
Thyme 4																	
Bailey 2,3,4																	
Anne 4																	

[1] See table 3.2 for an explanation of symbols. Insufficient data on presence in 1986.

Table 3.4. **Outcome of Agonistic Interactions among Adult Females in Troop Three**[1]

	Subordinate										
Dominant	Bt	Gp	CC	HL	SLE	#14	#13	TU	WE	MD	BT
Birdy Toe 1,2,3,4		12	6	8	9	2	7	7	8	4	9
Gimpy 3,4			7	2	6	4	4	1	5	2	1
Cleft Chin 3,4				3	6	1	0	1	6	2	1
Hot Lips 1,2,3,4					3	0	0	2	6	1	2
Scraggle Left Ear 3,4				1[2]		3	3	2	4	2	2
#14 Mark 2,4							4	2	3	0	3
#13 Notch Ear 4								1	6	2	3
Toes Up 1,2,3,4									3	3	5
White Eye 1,2,3,4										5	1
Miss Digit 1,2,3,4											1
Bend Tail 1?,2,3,4											

[1] The value in each cell is the frequency that the dominant won an agonistic bout with the subordinate. The numbers after the name in the first row refer to the years that the animal was present: 1 = 1986; 2 = 1990; 3 = 1991; 4 = 1992. X = not present because of death or age (too young to be an adult).

[2] Scraggle Left Ear bites Hot Lips, who took her banana.

to Toes Up, who grooms her on the back. Likewise, a subordinate female may signal a dominant by grinning while she approaches prior to affiliation. Such behaviors are unidirectional, although they can be ignored.

Adult males also exhibit these behaviors toward adult females as well as toward other males. Two Fingers, for example, would raise his eyebrows or lip smack at females, who would then approach and groom him.

Such a dominance hierarchy has been termed despotic, and it is similar to other species of macaques such as *M. mulatta* and *M. fuscata*. *M. fascicularis* adult males in multimale troops also have strict linear hierarchies in the wild (Wheatley, 1982).

Coalitions and Appeal Aggression

Another interesting set of behaviors involves multiple individuals, usually triads. These behaviors have been called *appeal aggression* because they are used by one individual toward another while soliciting aid from a third individual (De Waal, Van Hoof, and Netto, 1976). These behaviors are: (1) protected threat, in which an individual threatens another and presents his hindquarters to a third individual; (2) show-looking, in which an individual repeatedly threatens another while rapidly glancing back and forth at a third individual in a nonthreatening way; and (3) scream-threat or bark-screaming, in which an individual, usually young, screams at or near an individual, often while showing both aggression and submission behaviors in succession.

Some of the most exciting and interesting observations involve appeal aggression. Except for a brief comment by Shirek-Ellefson (1972) on scream-threats, these behaviors have not been reported in detail in the literature on any free-ranging population of *M. fascicularis*. Most field research does not have such a good degree of habituation of the animals and individual recognition as this site. Even under the best conditions, however, it is not always easy to see everything. Minimally, of course, three individuals are involved, but often as many as a half-dozen or more can be involved. In some cases, it is difficult to determine the target and the recipient of the behavior, especially because it occurs so quickly and is often chaotic. When multiple types of appeals are given, however, the situation is easily clarified, such as when an adult female scream-threats another individual, called the target, usually male, and show-looks a recipient, again usually a male, for assistance.

Data from more than 200 episodes of appeal aggression during the summers of 1986, 1990, 1991, and 1992 show two very clear patterns. The first and most obvious pattern is that adult females give over 90 percent of the appeal aggression, especially the scream-threats, which are usually higher-intensity threats than the show-looking and protected threats. Of course, there are typically two to three times more adult females than adult males in a troop, but each episode of appeal aggression often had multiple female responses, especially in the scream-threats. The second obvious pattern is that males are the target of these triadic interactions. In 98 percent of the scream-threat episodes, for example, females are soliciting other individuals for support against males. Only 2 percent of the time do females solicit individuals to support them against females. This pattern was also seen in show-looking and in protected threats. In addition, adult males are usually the recipient of these interactions. That is, males rather than females are more than twice as likely to be appealed to for assistance in aggressive interactions. Again, this was the case in scream-threats, show-looking, and protected threats.

Perhaps a few examples will better illustrate what happens. On the morning of June 26, 1992, the alpha female of Troop One named Tubby (see Plate 3.3) was up in a tree being groomed by an adult male named Scarface. After a half hour, she climbed to the ground and lay down next to five adult females, Spot Belly, White Eye, Dark Eye, Eyelid, and Lefty. Eyelid then proceeded to groom Tubby. Several other females approached, including Point Right Ear, a young juvenile female, and two infants, a male infant and a female infant. The alpha male, Two Fingers, then came up to Tubby and bit her. She gave scream threats to Two Fingers and fled. Five females then chased Two Fingers, and Tubby joined the chase. A few minutes after the threats and chasing, Two Fingers came up to Tubby and groomed her in reconciliation.

Another example is somewhat typical in sexual situations. Adult female Black Mandible in Troop One was by herself on the afternoon of August 4, 1992. A subadult male approached her and began to copulate with her, but before he had ejaculated, she ran off, giving scream-threats, and she then show-looked to the alpha male of her troop, Two Fingers.

Threats are often seen when contact aggression occurs either on an infant or on an adult female by an adult male, or after a non-resident subadult or adult male copulates in the vicinity of a resident adult male. Adult males will also scream-threat and show-look other males, as occurred, for example, in 1990 in Troop One, when Cut Tail male gave scream-threats to the beta male and then show-looked to the alpha male,

Plate 3.3. *The dangers of overfed animals are obvious in this animal, known by the name of Tubby.*

Two Fingers, after the beta male copulated with the alpha female.

Adult males use appeal aggression to form coalitions against other adult males. For example, in 1991, in Troop One, the fourth-ranking male, Scarface, scream-threats Black Ball male, a nonresident adult male, who has just copulated with the Troop One alpha female. Two Fingers, the Troop One alpha male, then helps him against Black Ball.

Animals learn appeal aggression as infants. One episode in 1986 in Troop One was rather funny. An old infant male of White Eye (a high-ranking adult female) gave scream-threats to a young juvenile male. He then ran screaming up to the juvenile and plopped down in fetal position with his head turned up, screaming right at the juvenile's feet. Seeing that no one was supporting the infant, the juvenile sat on the infant. Again, the juvenile looked all around, and when he saw no one coming to the aid of the infant, he started jumping up and down on the infant. Eventually support came, and the infant got up and chased and bit the juvenile. This last example also shows the severity of the outcome of appeal aggression. Animals can get seriously bitten, something Shirek-Ellefson (1972) did not see in her study on this species in the Botanical Gardens in Singapore.

Not all cases of appeal aggression involve males. On July 30, 1992, Spot Belly ranked number 12 in Troop One, threatened Point Right Ear, the lowest-ranking female in the troop, and then show-looked to Slappy, the third-ranked female. Spot Belly threatened Point Right Ear again, and Slappy also threatened Point Right Ear. Slappy then show-looked to Spot Belly, who groomed Slappy in return. Point Left Ear, another Troop One female, then approached and groomed Slappy. Spot Belly then threatened and chased Point Right Ear, the lowest-ranking female in Troop One. Another case involving show-looking to a female is also interesting because it involves a mother appealing to who we think is her daughter and an eventual reconciliation of the conflict. This episode occurred on August 5, 1992, and involved Slappy, the third-ranked female, who had just abandoned her dead newborn male the day before (this was her first infant). Slappy approached Point Right Ear, the lowest-ranking female in the troop, and her old infant male, by doing an eyebrow raise that was reciprocated by Point Right Ear. A subadult male, Mark Tail, approached and displaced Point Right Ear, who moved off one meter. Slappy then went and sat next to Point Right Ear. Two minutes later, White Eye, ranked number two in the troop and probably Slappy's mother, approached. Soon afterward, the infant of Point Right Ear geckered (a loud, staccato call—see "Vocalizations" later in the chapter), and White Eye chased and made contact aggression on the infant twice. The infant's mother was then chased after White Eye show-looked Slappy. The mother, Point Right Ear, fled from the two of them forty meters and up a tree, where she grinned and lip smacked at the two females. Both Slappy and White Eye then climbed up the tree, both of them giving pout lip smacks in a reconciliation.

Sometimes appeal aggression is used over food, as when the alpha male of Troop One, Two Fingers, took White Eye's (adult female) coconut away, and she gave scream-threats. A more hilarious and interesting case because it could be a case of lying occurred in 1991 when Birdy Toe (alpha female of Troop Three) alternated between scream-threatening and show-looking to everyone around her while she ate four to five bananas from a tourist as fast as she could.

Females will also solicit other females for aid, although not as often as they solicit males for aid. In other cases, it is not clear from whom exactly they are soliciting aid. It could be an appeal for help from anyone. These types of cases often involve large numbers of individuals, sometimes including the whole troop. The result is a sort of rapid mobilization force, which can be quite overwhelming. Large numbers of females can be rapidly massed to counter male aggression, for example, by the alpha male on the alpha female or on an infant. For example, in 1991, the alpha male of Troop Three, Line Nose, was quickly surrounded by four adult females, including Gimpy and Birdy Toe (the first- and second-ranking females), who gave him scream-threats after he bit someone. The alpha male of Troop One, Two Fingers, once pinned and held a brown infant down on the ground. He glanced all around after the infant squealed, as if anticipating the eventual mass attack of about seven individuals who quickly threatened him upon hearing the squealing. He then let the infant go. Sometimes the alpha male will actually be chased and bitten by adult females of the troop.

This mobbing effect is also used on potential predators. I saw it used against dogs, tourists, and myself. Most dogs are no threat to monkeys. I have seen monkeys play with dogs, monkeys grooming dogs, monkeys presenting to dogs, and I have even seen small infants playing together less than a meter away from a dog. In the latter example, the dog even licked the rear end of one of the infants. Certain dogs seem to be recognized by monkeys as being friendly or not. I have never seen dogs hurt, much less kill, a monkey, and several local people who frequented the Monkey Forest told me they have never heard of a dog killing a monkey. Nevertheless, some monkeys were definitely afraid of certain dogs and would quickly climb a tree when encountering them. Dogs and monkeys also competed for food at the graveyard, and the dogs would be attacked by individual adult males or mobbed by animals, in some cases only females scream-threatening. Tourists and researchers could also get mobbed instantly by coalitions of animals, especially when infants screamed after being touched. I saw one tourist get badly bitten on the arm after he moved too close to a dying infant; he was mobbed and jumped on by many animals from the troop. On June 10, 1991, I saw Troop One's alpha male Two Fingers and as many as eight other animals attack and bite a tourist when he grabbed an infant. I also never saw or heard monkeys afraid of any raptors or alarm-call to them. Young infants would play in muddy rice fields, for example, while a serpent eagle circled and called overhead.

There was also a very interesting story told to me by a local villager and confirmed by another farmer. They witnessed an attack by many monkeys on a python, biting and killing it after it caught an infant. After being attacked by the monkeys, the snake fell down into the gorge. When the villagers went to retrieve the python skin, they found it of no use because it was too badly bitten up. The infant monkey was also gone. This is another example of how monkeys are enemies of snakes (see chapter 1). I have seen monkeys alarm-call, slap at, and mob snakes. I saw a dead civet cat in the graveyard in 1992, but what danger this cat might pose to monkeys is unknown.

Appeal aggression is also used extensively in mobilizing the troop against other troops. An interesting case occurred on July 18, 1991. Troop One chased Troop Three from the graveyard, where they were feeding, to the first lookout on the other side of the gorge. During this aggressive encounter, the alpha male of Troop Three bit the beta male of Troop One on the side, but no blood was apparent to the naked eye. The beta male of Troop One then mounted Bent Toe, the beta male of Troop Three, who then threatened his own troop member, Curl Ear, a subadult male, several times within the span of one minute. The alpha male of Troop Three, Line Nose, was chased up a tree, but he came down to support his own adult females who were scream-threatening Troop One. He then ran up the tree again while Bent Toe threatened Curl Ear up a tree. Curl Ear came down the tree, and Bent Toe did a pout face to him. Scraggle Left Ear (adult female) of Troop Three then presented to Curl Ear, and the latter groomed her. She then groomed him while they hid at the first lookout. Bent Toe lip smacked and groomed Umbilical Boy, a subadult male from Troop Two, who had just attacked Bent Toe's troop. Rather than assist his own troop fight Troop One, Bent Toe, an older male, seemed to switch sides temporarily.

Grooming and Rank

Females

The first obvious pattern in adult female grooming is that lower-ranking females tend to groom higher-ranking females (Table 3.5). An examination of over 1,000 grooming sessions in the three troops in the years 1991 and 1992 reveals that 72 percent of all grooming sessions consisted of lower-ranking females grooming higher-ranking females. This pattern was approximately the same in each year for each troop.

An analysis of female intratroop grooming sessions in Troop One is presented in Table 3.6. There were 269 focal samples on Troop One adult females ($N = 14$) for 1992, each of which was fifteen minutes long. Again, higher-ranking adult females tend to receive more grooming than lower-ranking females. In 1992, for example, there were a total of 126 grooming bouts (68 percent of all bouts observed) up the female dominance hierarchy and 58 grooming bouts (32 percent of all bouts observed) down the hierarchy (Table 3.6). The second pattern is that more dominant females receive longer durations of grooming than they give (see Tables 3.7, 3.8, 3.9, 3.10). This pattern is reversed for the more subordinate females. For example, in Troop One in 1992 (Table 3.8) we can compare the grooming duration of the top seven-ranked females with those of the bottom seven-ranked females. The higher-ranked individuals received a total of 477 minutes of grooming (73 percent) out of 657 minutes while only giving a total of 294 minutes (44 percent) out of 667 minutes of grooming. The lower-ranked individuals, however, only received a total of 180 minutes of grooming (27 percent) out of the 657 minutes while giving a total of 373 minutes (56 percent) out of the 667 minutes.

Table 3.5. **Total Number of Adult Female Grooming Sessions for All Three Troops in 1991 and 1992**[1]

Year	Troop	Tot. Focals	# Tot. Grm. Sess.	#High on Low		#Low on High	
1991	1	274	160	49	31%	111	69%
1992	1	269	184	58	32%	126	68%
1991	2	255	140	43	31%	97	69%
1992	2	527	317	75	24%	242	76%
1991	3	213	46	12	26%	34	74%
1992	3	358	160	44	27.5%	116	72.5%
Both	all	1,896	1,007	281	28%	726	72%

[1] The percentages of grooming sessions are given in which higher-ranking females groomed lower-ranking females and lower-ranking females groomed higher-ranking females. The data are obtained from the focal samples taken each year on the adult females. A grooming session is defined as any grooming bout between individuals lasting 30 seconds or longer. Mutual grooming was rare (never among adult females during this observation time).

Table 3.6. **Total Number of Grooming Sessions (30 seconds or longer) Observed in 1992 among Troop One Adult Females**

Groomers (vertical row) — **Adult Females Receiving the Sessions (horizontal row)**

Adult Females by rank[1]	**e? Tb**	**om WE**	**p Sl**	**m BM**	**e Y15**	**m Ey**	**m St**	**e Y8**	**m DE**	**om Lf**	**e PLE**	**e SB**	**e TI**	**m PRE**	**Total Given**
Tubby[2]				13		4	3		1			2		1	24
White Eye	2		4	2								1			9
Slappy	1	7		1				1	1				1		12
Black Mandible	3		1			1					2[3]		1		8
Young #15	2	3		5		2	1		1						14
Eyelid	5	2		4	1										12
Scartail	3	1		3	1	2			2		1			1	14
Young #8			1	1		2	3								7
Dark Eye						3	2			3		1			9
Lefty	1	1		3		2	6					1			14
Point Left Ear	1	1		8	1			2	1	2				1	17
Spot Belly	2			5	1	2	7		2	4				2	25
Toe In	1		1				2	1	3	1				3	12
Point Right Ear				1		3			1		1		1		7
Total = number of sessions received	**21**	**15**	**7**	**46**	**4**	**21**	**24**	**4**	**12**	**10**	**4**	**5**	**2**	**9**	**184**

[1] Reproductive State: e = estrous; om = mother of infant newly weaned, still seen occasionally on nipple; m = mother of nursing infant; p = pregnant.

[2] Tubby only had two sessions on another AF (Spot Belly) without an infant present. In all other cases, she groomed a mother with the infant on the mother at the time.

[3] Black Mandible had two sessions on Point Left Ear when Point Left Ear was holding her young brown infant.

Table 3.7. **Total Number of Minutes Spent by Troop One Adult Females Grooming and Being Groomed by Other Adults in the Troop, 1991**[1]

Rank	AF	Cycle	Grm w/AF			Grm w/AM			Total
			On	By	Tot.	On	By	Tot.	
1	Alpha	e	19	70	89	39	41	80	169
2	Tubby	n	30	54	84	78	3	81	165
3	White Eye	p	33	47	80	20	1	21	101
4	Black Mandible	m	53	113	166	18	0	18	184
5	Eyelid	m/md	38	58	96	7	0	7	103
6	Scar Tail	md	38	54	92	12	0	12	104
7	Dark Eye	md/e[2]	67	35	102	60	32	92	194
8	Lefty	e	40	23	63	25	30	55	118
9	Point Left Ear	e	69	3	72	3	13	16	88
10	Spot Belly	m/md	63	47	110	6	1	7	117
11	Toe In	e	40	0	40	6	82	88	128
12	Point Right Ear	m	36	22	58	0	0	0	58

[1] From the 1991 focal animal data. AF = Adult Females; AM=Adult Males; Grm = Groom; e = estrous; om = mother of older infant, seen nursing occasionally; m = mother of nursing infant; md = mother of infant that dies during study, not yet in estrus; p = pregnant, n = anestrous

[2] Dark Eye came into estrus a few weeks after losing her infant.

Table 3.8. **Total Number of Minutes Spent by Troop One Adult Females, Grooming and Being Groomed by Other Adults in the Troop, 1992**[1]

Rank	AF	Cycle	Grm w/AF			Grm w/AM			Total
			On	By	Tot.	On	By	Tot.	
1	Tubby	e[2]	51	74	125	17	24	41	166
2	White Eye	om	23	47	70	18	36	54	124
3	Slappy	p	43	29	72	0	0	0	72
4	Black Mandible	m	23	139	162	6	0	6	168
5	Young #15	e	44	11	55	14	10	24	79
6	Eyelid	m	55	89	144	39	1	40	184
7	Scar Tail	m	55	88	143	9	1	10	153
8	Young #8	e	22	21	43	0	0	0	43
9	Dark Eye	m	46	51	97	7	10	17	114
10	Lefty	om	44	44	88	39	1	40	128
11	Point Left Ear	e	63	12	75	11	38	49	124
12	Spot Belly	e	98	12	110	18	8	26	136
13	Toe In	e	58	9	67	21	37	58	125
14	Point Right Ear	m	42	31	73	1	0	1	74

[1] From the 1992 focal animal data. See Table 3.7, above, for explanation of symbols.

[2] Tubby was copulating with alpha and other troop males in 1992 although not seen to copulate in the previous years. She is presumed sterile, having short nipples.

Table 3.9. **Total Number of Minutes Spent by Troop Two Adult Females Grooming and Being Groomed by Other Adults in the Troop, 1991**[1]

Rank	AF	Cycle	Grm w/AF			Grm w/AM			Total
			On	By	Tot.	On	By	Tot.	
1	PD	m	17	70	87	5	0	5	92
2	Guynan	m	43	32	75	0	0	0	75
3	Jordi	m	24	83	107	0	0	0	107
4	Kidnapper	sterile	29	37	66	0	0	0	66
5	Abby	e	33	18	51	6	13	19	70
6	Scarface	m	40	73	113	0	0	0	113
7	Daisy	m	31	26	57	0	0	0	57
8	Worf	e	33	12	45	15	1	16	61
9	Humpback	e	22	2	24	0	0	0	24
10.5	Gimpy	m	10	31	41	0	0	0	41
10.5	Beverly	p-nb	78	35	113	0	0	0	113
12	Maggie	e	22	9	31	14	25	39	70
13	Ada	p-nb	19	5	24	0	0	0	24
14	Bailey	e	20	3	23	1	0	1	24

[1] From the 1991 focal animal data. p-nb = pregnant, newborn; sterile = no estrus, no infant ever observed. See Table 3.7 for explanation of additional symbols.

Table 3.10. **Total Number of Minutes Spent by Troop Two Adult Females Grooming and Being Groomed by Other Adults in the Troop, 1992**[1]

Rank	AF	Cycle	Grm w/AF			Grm w/AM			Total
			On	By	Tot.	On	By	Tot.	
1	PD	m	29	163	192	23	0	23	215
2	Juniper	m	37	86	123	8	0	8	131
3	Guynan	m	22	126	148	0	0	0	148
4	Jordi	e	18	33	51	17	11	28	79
5	Kidnapper	sterile	43	48	91	7	0	7	98
6	Abby	e	33	18	51	47	3	50	101
7	Scarface	e	76	52	128	10	92	102	230
8	Daisy	p?	51	55	106	0	0	0	106
9	Worf	e	81	54	135	8	110	118	253
10	Humpback	p-nb	40	61	101	0	0	0	101
11.5	Beverly	m	124	71	195	0	0	0	195
11.5	Gimpy	e	15	1	16	12	2	14	30
13	Maggie	e	72	6	78	2	19	21	99
14	Ada	p?	71	5	76	2	0	2	78
16	Thyme	e	41	24	65	10	61	71	136
16	Bailey	e	34	0	34	41	33	74	108
16	Anne	e	18	1	19	3	17	20	39

[1] From the 1992 focal animal data. See above for explanation of symbols.

Table 3.11. **Total Number of Minutes Spent by Troop Three Adult Females Grooming and Being Groomed by Other Adults in the Troop, 1991**[1]

Rank	AF	Cycle	Grm w/AF			Grm w/AM			Total
			On	By	Tot.	On	By	Tot.	
1	Birdy Toe	m	2	66	68	74	1	75	143
2	Gimpy	m	12	23	35	17	0	17	52
3	Cleft Chin	e	27	12	39	2	0	2	41
4	Hot Lips	m	22	24	46	1	0	1	47
5	Scraggle Left Ear	m	35	19	54	20	0	20	74
6	Toes Up	e	41	12	53	3	2	5	58
7	White Eye	e	19	11	30	0	0	0	30
8	Miss Digit	p-nb	4	2	6	12	0	12	18
9	Bend Tail	e	7	0	7	26	7	33	40

[1] From the 1991 focal animal data. AF = Adult Female; AM = Adult Male. Cycle: e = estrous; m = mother of nursing infant; p = pregnant; p-nb = pregnant, newborn; sterile = no estrus, no infant ever observed.

Table 3.12. **Total Number of Minutes Spent by Troop Three Adult Females Grooming and Being Groomed by Other Adults in the Troop, 1992**[1]

Rank	AF	Cycle	Grm w/AF			Grm w/AM			Total
			On	By	Tot.	On	By	Tot.	
1	Birdy Toe	m	6	79	85	26	6	32	117
2	Gimpy	m	49	34	83	10	0	10	93
3	Cleft Chin	m	28	54	82	2	0	2	84
4	Hot Lips	m	25	57	82	11	3	14	96
5	Scraggle Left Ear	m	67	58	125	30	0	30	155
6	#14 Mark 2	p?	26	34	60	24	0	24	84
7	Notch Ear	p-nb	51	38	89	3	0	3	92
8	Toes Up	e	39	13	52	20	70	90	142
9	White Eye	e	23	0	23	15	65	80	103
10	Miss Digit	e	40	0	40	28	31	59	99
11	Bend Tail	e	17	4	21	17	0	17	38

[1] From the 1992 focal animal data. See above for explanation of symbols.

Females thus tend to groom up the hierarchy, and this may help them to build coalitions with more dominant animals. This hypothesis is called the affiliation for support hypothesis (Seyfarth, 1977), and it predicts that higher-ranking females interfere with the affiliative attempts of low rankers toward high rankers. Our data also support this. In some of the larger groups of high-ranking and middle-ranking females who huddle and groom each other, for example, middle-ranking females in the group will displace, chase, and sometimes contact aggress lower-ranking females who try to join them. Females will also use their dominance to interfere with other females who are with the alpha male. In 1992, for example, White Eye, the third-ranked female in Troop One, chased Dark Eye, ranked number nine, away from the alpha male Two Fingers when she saw Dark Eye grooming him. White Eye then approached him, and they alternated grooming each other. One unusual aspect about grooming in these monkeys is that mutual grooming is exceedingly rare. It was only seen a couple of times during our entire study period. When it occurred, the more dominant animal usually slapped the other's hand away and continued grooming or just stopped grooming.

To better understand a female's behavior, however, we need to examine her reproductive condition and kinship connections. Three major reproductive variables are important—motherhood, estrus, and pregnancy. Tables 3.7 through 3.12 give these cycle states and the number of minutes of grooming for every adult female in all three troops for the years 1991 and 1992. One of the most notable effects on a troop's females is the presence of infants. Infants and, of course, their mothers are the objects of intense curiosity and attention. These tables show that mothers are some of the most frequently groomed and longest-groomed females in the troop. Black Mandible of Troop One, for example, was the mother of the youngest brown infant (a female) in 1992. She was the most frequently groomed individual in the troop that year by females (see Tables 3.6, 3.7, 3.8 and Figure 3.1). She was not only groomed twice as frequently as the alpha female, but also more than the alpha male. She also received more total minutes of grooming than any other female in the troop that year. Two other mothers in the troop, Eyelid and Scartail, received as many grooming bouts as the alpha female (see Table 3.6). These tables and Figure 3.1 also show that estrous females receive very little grooming from other females. Tubby is unusual in that she received considerably more grooming than any other estrous female in 1992, but this is probably because she is the alpha female. Her estrous condition is unusual, maybe even suspect, because she was never observed to copulate before this year. Another way to compare behavior between reproductive states is to examine the same individual who is in estrus one year and a mother in another year. There are three such individuals: Cleft Chin in Troop Three and Jordi and Gimpy in Troop Two. When these three females were in estrus, adult females groomed them less than they groomed other females. This situation was reversed, however, when the three individuals had an infant.

Estrus does not appear to dramatically affect the adult female grooming of adult males, especially to the extent that it affects male grooming behavior. Estrous and nonestrous females are all grooming adult males although females in estrus seem to groom males more. One confounding variable here is that the alpha females of all three troops appear to have a special relationship with the alpha males. The alpha female grooms the alpha male a lot regardless of her cycle state. One pattern that does appear in all three troops during 1991 and 1992 is that the alpha male received the most grooming sessions (Table 3.13).

Table 3.13. **Adult Females' (AF) Grooming Sessions on Adult Males Observed within the Three Troops, 1991–1992**

Troop One—1991

Adult Male	Rank	By Estrous AF (*N* = 5)	By Nonestrous AF (*N* = 7)	Total
Two Fingers	1	9	17	26
Crooked Tail	2	4	5	9
Cut Tail	3	10	5	15
Scarface	4	5	3	8
Total		**28**	**30**	**58**

Troop One—1992

Adult Male	Rank	By Estrous AF (*N* = 6)[1]	By Nonestrous AF (*N* = 8)	Total
Two Fingers	1	13	28	41
Cut Tail	2	7	4	11
Scarface	3	5	7	12
Total		**25**	**39**	**64**

Troop Two—1991

Adult Male	Rank	By Estrous AF (*N* = 5)	By Nonestrous AF (*N* = 9)	Total
Arnold	1	9	1	10
Tommy	2	3	0	3
Pointer	3	1	0	1
Total		**13**	**1**	**14**

Troop Two—1992

Adult Male	Rank	By Estrous AF (*N* = 9)	By Nonestrous AF (*N* = 8)	Total
Arnold	1	8	13	21
Tommy	2	12	4	16
Pointer	3	5	0	5
Martin	4	4	0	4
Total		**29**	**17**	**46**

Troop Three—1991

Adult Male	Rank	By Estrous AF (*N* = 4)	By Nonestrous AF (*N* = 5)	Total
Line Nose	1	8	20	28
Bent Toe	2	0	11	11
Curl Ear	3	4	1	5
Total		**12**	**32**	**44**

Troop Three—1992

Adult Male	Rank	By Estrous AF ($N = 4$)	By Nonestrous AF ($N = 7$)	Total
Line Nose	1	13	25	38
Bent Toe	2	4	8	12
Curl Ear	3	0	0	0
Young Male	4	5	1	6
Total		**22**	**34**	**56**

Total all three troops 1991–1992: 282 sessions—by estrous AF = 129 (46%)
by nonestrous AF = 153 (54%)

[1] includes Tubby, copulating in 1992, never seen to copulate previously.

Pregnancy may also have an effect. Slappy, the number three-ranked female, for example, also received very little grooming while she was pregnant. Kinship is also an important factor, of course, because mothers and daughters tend to groom each other. Slappy and her mother, White Eye, for example, groomed each other more than they groomed any other female (see Table 3.6).

These reproductive and kinship variables obviously have a major effect on female behaviors such as grooming. Nevertheless, while it is the case that some animals affiliate with high rankers in order to obtain their agonistic support, it may also be the case that such affiliation is pleasurable and that a sense of troop cohesion is built. For example, Table 3.6 shows that the lowest-ranking female, Point Right Ear, actually received more grooming bouts than six other higher-ranking females, and that she was groomed by Tubby, the alpha female. These observations show the essential cohesiveness of the troop, which is useful in troop-troop competition. In all of these ways intratroop competition can be mitigated and allow the troop to remain cohesive and large for purposes of intertroop competition.

Males

An examination of male grooming sessions on adult females shows several patterns. The first and most obvious pattern is that in all three troops the males tend to groom estrous females rather than nonestrous females (Table 3.14). Table 3.14 shows that 88 percent of all male grooming sessions were by males on estrous females whereas only 12 percent of their sessions were on nonestrous females ($N = 223$). The second pattern is that the alpha male tends to groom the most, doing two-thirds of the entire grooming (Table 3.14).

The grooming patterns of the adult males in Troop One reveal their integration with the troop's adult females (see Table 3.15). This table shows that Two Fingers in Troop One received more sessions of grooming than any other individual in Troop One, with the exception of Black Mandible, who had the youngest baby. He also groomed and was groomed by almost all of the troop's adult females. The other two adult males were groomed by less than half of the troop's adult females, and most of these females were of low rank.

Table 3.14. **Adult Males' (AM) Grooming Sessions on Adult Females Observed within the Three Troops in 1991 and 1992**

Troop One—1991

AF Cycle	N	Tot. # Grm. Sessions	AM1	AM2	AM3	AM4
Estrous AF	5	40	29	3	4	4
Nonestrous AF	7	3	2	0	1	0
Total	**12**	**43**	**31**	**3**	**5**	**4**

Troop One—1992

AF Cycle	N	Tot. # Grm. Sessions	AM1	AM2	AM3
Estrous AF	6[1]	27	17	6	4
Nonestrous AF	8	19	16	0	3
Total	**14**	**46**	**33**	**6**	**7**

[1] includes Tubby, previously never seen in copulation, now copulating.

Troop Two—1991

AF Cycle	N	Tot. # Grm. Sessions	AM1	AM2	AM3
Estrous AF	5	11	9	1	1
Nonestrous AF	9	0	0	0	0
Total	**14**	**11**	**9**	**1**	**1**

Troop Two—1992

AF Cycle	N	Tot. # Grm. Sessions	AM1	AM2	AM3	AM4
Estrous AF	9	77	56	17	1	3
Nonestrous AF	8	0	0	0	0	0
Total	**17**	**77**	**56**	**17**	**1**	**3**

Troop Three—1991

AF Cycle	N	Tot. # Grm. Sessions	AM1	AM2	AM3
Estrous AF	4	5	4	1	0
Nonestrous AF	5	1	0	1	0
Total	**9**	**6**	**4**	**2**	**0**

Troop Three—1992

AF Cycle	N	Tot. # Grm. Sessions	AM1	AM2	AM3	AM4
Estrous AF	4	38	13	20	2	3
Nonestrous AF	7	3	3	0	0	0
Total	**11**	**41**	**16**	**20**	**2**	**3**

Total all 3 troops 1991–92: 223 sessions—197 on estrous AF (88%)
26 on nonestrous AF (12%)

Table 3.15. **Total Number of 1992 Grooming Sessions in Which Troop One Adult Males (AM) (*N* = 3) Are Groomed by Adult Females of Troop One (*N* = 14)**[1]

			Adult Males by Rank			
AF	AF Rank	Cycle	#1	#2	#3	Total
Eyelid	6	m	12	2	1	15
Toe In	13	e	7	0	1	8
White Eye	2	om	7	0	0	7
Lefty	10	om	2	2	3	7
Point Left Ear	11	e	0	5	0	5
Spot Belly	12	e	1	0	4	5
Scartail	7	m	3	0	1	4
Young #15	5	e	2	2	0	4
Tubby	1	e	3	0	0	3
Dark Eye	9	m	1	0	2	3
Black Mandible	4	m	2	0	0	2
Point Right Ear	14	m	1	0	0	1
Young #8	8	e	0	0	0	0
Slappy	3	p	0	0	0	0
Total			**41**	**11**	**12**	**64**

[1] See Table 3.7 for explanation of symbols.

Fifty-seven focal samples, each fifteen minutes in length, were taken on the three adult males in Troop One in 1992. Grooming sessions between males are virtually nonexistent. Despite the multimale troop structure, adult males in the three troops hardly ever groomed each other (Table 3.16).

Infant Handling and Lethal Kidnapping

The birth of an infant and its survival is essential to the reproductive success of a female. The death of an infant, therefore, has a serious effect on a mother's reproductive success. Gestation length is reported to be about 165 days in *M. fascicularis* (Jewett & Dukelow, 1972) so that the loss of a newborn precludes any reproductive success to a female for that year.

We have seen more infant deaths than any other age-sex class. Twelve infants (brown and black) died during our four summers of research. Newborn or black infants are the most vulnerable. Out of the twelve newborns observed, eight survived. The older, or brown, infants appeared to have a much better chance of survival. They are pulled and grabbed at and temporarily taken for short periods of time, but they eventually get back to their mothers on their own. Out of a total of fifty-four brown infants, all but three survived during our

Table 3.16. **Adult Male Grooming or Groomed by Another Adult Male in the Same Troop**[1]

Troop One—1991

Adult Male (groomer)	Rank	Two Fingers	Crooked Tail	Cut Tail	Scarface
Two Fingers	1	—	1 (10)	0	0
Crooked Tail	2	0	—	0	0
Cut Tail	3	0	0	—	0
Scarface	4	0	0	0	—
Total		**0**	**1 (10)**	**0**	**0**

Troop One—1992

Adult Male	Rank	Two Fingers	Cut Tail	Scarface
Two Fingers	1	—	0	0
Cut Tail	2	0	—	0
Scarface	3	1 (1)	0	—
Total		**1 (1)**	**0**	**0**

Troop Two—1991

Adult Male	Rank	Arnold	Tommy	Pointer
Arnold	1	—	0	0
Tommy	2	0	—	0
Pointer	3	0	0	—
Total		**0**	**0**	**0**

Troop Two—1992

Adult Male	Rank	Arnold	Tommy	Pointer	Martin
Arnold	1	—	0	0	0
Tommy	2	0	—	0	0
Pointer	3	0	0	—	0
Martin	4	—	2(3)(2)	0	—
Total		**0**	**2(3)(2)**	**0**	**0**

Troop Three—1991

Adult Male	Rank	Line Nose	Bent Toe	Curl Ear
Line Nose	1	—	0	0
Bent Toe	2	0	—	0
Curl Ear	3	0	0	—
Total		**0**	**0**	**0**

Troop Three—1992

Adult Male	Rank	Line Nose	Bent Toe	Curl Ear	Young Male
Line Nose	1	—	0	0	0
Bent Toe	2	0	—	0	0
Curl Ear	3	0	0	—	2(1)(4)
Young Male	4	0	0	3(1)(1)(4)	—
Total		**0**	**0**	**3(1)(1)(4)**	**2(1)(4)**

[1] Number of sessions on or by AM with minutes of each session in parentheses.

summer research sessions. Mothers will carry dead infants around for days, grooming and cradling the body. It is interesting that some Balinese living right next to the Monkey Forest had never seen a monkey corpse by itself and spoke about it as a sort of religious mystery. I have seen mothers finally abandon their dead infants in the evening; by the next morning, the bodies are gone. Once a mother abandoned her dead infant just before noon, and after noting where it was lying, I went away for two hours and returned to find the corpse missing, probably eaten by dogs. Once I saw a dog unsuccessfully try to get a dead infant away from the mother by tugging on it.

Probably the most important event for a newborn is that it nurses within the first few days of life. We have seen four infants remain with their mothers but die because they were never seen to obtain milk from the mother. Three of these mothers were first-time mothers in Troop One. Scar Tail, for example, lost a newborn male in 1991 (her infant of 1992 survived). Spot Belly also lost a newborn female that same year. Slappy lost a newborn male in 1992. None of these first-time mothers had elongated nipples, by which we judged an adult female as being multiparous. Nor did any of the first-time mothers have enlarged breasts, which would indicate the presence of an adequate supply of milk.

Three newborns, or black infants, and three brown infants died from injuries. In 1986, a newborn female and an older brown infant disappeared from Troop One, and both their mothers had fresh wounds at the time of their disappearance. In 1990, a newborn male in Troop Two died from a massive wound on his back. In 1991, Eyelid's young brown male infant died after suffering a massive skull injury. His mother, holding him ventrally, may have inadvertently smashed him on a limb or on the ground after a leap. Eyelid's arm was also seriously injured. That same year, Dark Eye's newborn male also disappeared. A guard said that he had seen it being "ripped apart." Both of these mothers were experienced, having been seen with infants in previous summers. In Troop Two, a first-time mother named Humpback was seen dragging her newborn female by the tail, holding the baby upside down, and sitting on her for the first day. The baby started nursing the next day and lived for three weeks, until it mysteriously disappeared.

Mothers of newborns are very protective of them, and they do not allow infants out of their grasp. Mothers of newborns attract great interest from other females, who attempt to touch, to groom, and to pull on them, but rarely do they succeed in taking the newborn from the mother. In some cases, however, the mother can lose the newborn, and when more dominant females end up with the newborn, the results can be lethal to the baby, though not always.

A Troop Three first-time middle-ranking mother named Notch-Ear lost her male newborn a day after his birth. The alpha and beta females, Birdy Toe and Gimpy, tugged and pulled at the baby and eventually kidnapped him. Both of these females had brown infants of their own whom they were still nursing. They proceeded to exchange the kidnapped baby between them, sometimes pushing their own infants away while holding on to the kidnapped newborn. At other times, they held him down with their feet while nursing their own babies.

The mother recovered her newborn male infant three and a half hours later when it slipped through the grasp of the foot of the female who had originally kidnapped it. This newborn was kidnapped again four days later by the alpha female Birdy Toe. An hour and thirteen minutes later, the mother spotted her baby on the ground amidst a group of adult females, and she charged in, grabbed him, and fled. The deaths of kidnapped newborns occur when higher-ranking and nonlactating females prevent the lower-ranking mothers from retrieving their babies. Kidnapped newborns last about three days before starving to death on nonlactating females.

We witnessed this event in August of 1990, just after our arrival at Ubud. On the morning of August 7, the third-ranked female, named Tubby, was seen with a black and pink (newly born) female. It is very unlikely that Tubby was the mother of this newborn. Tubby was seen on several occasions the day before without a newborn. The next morning there was no blood on Tubby or on the newborn. There was also no placenta or umbilical cord on the baby. There was no change in Tubby's tubbiness either, so the baby possibly was born on August 6. The baby was last seen with Tubby at 3:30 on August 9. Newly born infants usually are able to survive three days without food. It is unclear just where this pink-faced black-coated infant came from and who the mother was. It possibly could have come from a Troop One female. One possibility that we did not consider in 1990 was that the infant could have come from another troop. We witnessed this in 1991, when the alpha female of Troop One ended up with a young brown female infant from Troop Two, so it is also possible that Tubby somehow obtained this baby from another troop, possibly from Troop Two.

Together with Earthwatch volunteers, we continuously observed Tubby over the next few days. These observations show that Tubby was very protective of the baby. She tried to prevent other females, both adults and juveniles, from taking the baby by threatening them or moving off. Over an 8.5-hour period of time, an average of six individuals per hour approached Tubby, either to groom her or to touch the infant. Other individuals pulled or made a grab for the baby at a rate of almost ten per hour over a five- hour period. The next morning this rate had dropped to five per hour over a 3.5-hour period of time. Coalitions of females and the beta male constantly threatened and chased individuals away from the baby. The baby desperately tried to get food, and when she did, Tubby simply took it away. Tubby was never seen to nurse the baby, and it disappeared before the morning of the tenth, when Tubby was sighted again without the baby.

Another case of lethal kidnapping occurred later in 1990 in Troop Three. White Eye had a baby boy born the morning of September 2. She was still touching her perineum and licking her fingers at 9:24 that morning. An hour later a young juvenile and later Cleft Chin, an adolescent female, tried to grab the baby on two separate attempts. White Eye was last seen with her infant on her, huddling with Cleft Chin, at 11:00 the next morning. By 4:00 that afternoon, Cleft Chin was seen with the baby, sitting next to the adolescent, Gimp, and White Eye, the mother, who was being groomed by Miss Digit. Cleft Chin and Gimp ranked number three and number two, respectively, in the adult female hierarchy for the next year. White Eye made a grab for her infant but was unsuccessful in getting it back from Cleft Chin. The mother ran off to get food from tourists. The next morning, the top three ranked females threatened the mother away from her baby, although she did manage to groom it. Over the next few days, the mother was chased and bitten in her attempts to get near Cleft Chin and Gimp, who had her baby. The baby was never seen to nurse, and it was last seen on the afternoon of September 5.

In 1992 in Troop Two, the higher-ranking female, Kidnapper, kidnapped Humpback's black infant when it was less than three days old. The baby was observed to be close to death the last day it was seen. Kidnapper was very solicitous of the baby, cradling and grooming it, but the baby was dying from hunger. The mother made no attempt to retrieve the infant from the higher-ranked kidnapper during observation time.

One other kidnapping was very interesting because it was an intertroop kidnapping. In the morning of June 26, 1991, a juvenile in Troop One was seen holding a very young brown (approximately one month old) female infant. This infant was judged to come from Troop Two because all of the brown infants in Troop Three and Troop One were accounted for, having been observed before and after the event. Another Troop Two mother lost her newborn black infant at the same time and had a huge gash in her leg. It is possible that there was a troop-troop aggressive encounter during which the young brown female was separated from her mother and was snatched away by Troop One juveniles. Several Troop One juveniles held this infant during the following week and threatened off other interested adult females such as Point Right Ear and Spot Belly. About one week later, the alpha female who had just finished estrus, having been copulated with extensively, and who was not lactating, was seen holding this infant ventrally. The alpha female "adopted" this baby as she carried and protected it for the remainder of our time there and was still doing so when we left about a month later. The infant fed on tourist food, sometimes grabbing it from the alpha female. Observations on this infant were interesting because the alpha female never appeared to give the infant any of her food, even after the infant cooed. The alpha female also grabbed food away from the infant. The infant became very thin and appeared to get food only when the alpha female was full or indifferent to the infant taking it. Fortunately for the infant, the female was the alpha, thus having the best food supply possible to a female.

Lethal kidnapping is an extreme case of harassment. Mothers of infants and a pregnant female were seen to harass or abuse infants. White Eye, the second-ranked female in Troop One in 1992, appeared to go out of her way to abuse the young brown infant female of Black Mandible, the fourth-ranked female. White Eye bit, slapped, pulled its hair, pinned it to the ground, and kidnapped the baby twice for over an hour before she got away. Being outranked, Black Mandible did nothing to retrieve her baby; however, the infant never appeared to be injured by the abuse. Another case of abuse on infants was by Slappy, the third-ranked female in Troop One in 1992. A few weeks before she gave birth to her male infant, Slappy appeared to be extremely irritated by any infant in her vicinity and continually threatened and contact aggressed them.

Two other cases of harassment of infants are interesting because they show the importance of a female's reproductive condition in regard to rank and to other special types of relationships between individuals. The first case is that of the eleventh-ranked female, Point Left Ear, who was in Black Mandible's juvenile cohort during the late 1980s. Point Left Ear has never had a baby, judging by her nipple length, and she loved to handle brown infants. When Point Left Ear came into estrus in the summer of 1992, she kidnapped the young brown female infant of Black Mandible (the fourth-ranking adult female) on several occasions, upon which Black Mandible simply charged Point Left Ear and grabbed her baby back. One time, however, Point Left Ear ran and sat next to Two Fingers, the alpha male, who had been copulating with her throughout the day, and he threatened the mother away when she approached. Over the next thirty minutes, Black Mandible drew closer and closer very slowly to her baby on Point Left Ear. Finally, she was able to groom Point Left Ear, who hung onto the baby and grinned at Black Mandible. The mother then gradually pulled her baby back onto her and left.

The second case involves Black Mandible again and the fourteenth-ranked female, Point Right Ear. Both of these females are approximately the same size and age, having been born in 1986. They both had baby girls in 1991 who played together. One day, Point Right Ear's mother, Toe In, ranked number 13, threatened, chased, and generally harassed her granddaughter for over twenty minutes. Being outranked, Point Right Ear could not seem to protect her baby, even when next to her and even when her infant came to her to hide from Toe In. The baby finally ran up to and huddled with Black Mandible, and the harassment stopped because Toe In would not go close to Black Mandible. Eventually Point Right Ear, who never left the area and watched her infant and mother the entire time, got her baby back and nursed her.

Summary

Our Balinese data show that this species of macaque is similar to the other so-called despotic macaques. Both adult males and females have strict linear dominance hierarchies. Adult females especially have very stable and strong dominance hierarchies, as we have shown for thirty-five different adult females of three troops over three summers of research. The linear hierarchies of adult females in Troops One and Three include four years of research over a six-year period. In addition, we have several cases that confirm Kawamura's first principle that a daughter ranks just after her mother in the hierarchy for this species of macaque. These rank relationships are used to obtain certain resources such as food, and, for males, estrous females.

The behavior of adult females is also affected by their cycles. An estrous female is with adult and subadult males more than she is with adult females. She grooms less with adult females during this time. In some cases, a lower-ranked estrous female interacts very little with the other females and is more often to be found on the periphery of the troop, where a male might accompany her. A pregnant female tends to avoid males in contrast to estrous females. She becomes less social and appears to avoid conflict through her pregnancy, especially at the end. Mothers regardless of their rank are at the pinnacle of attractiveness to other females in the troop and are sought out and are also seeking out other mothers for grooming. Among wild troops as well, groups of adult females are significantly larger when mothers and infants are present (Wheatley, 1982).

We also have strong evidence for reproductive competition between females. Higher-ranking females harass and kidnap the infants of lower-ranking mothers. When these kidnappings are of newborns, they can be lethal. Interestingly, observations of lethal kidnapping have been reported in two of the other "despotic macaques," *M. mulatta* and *M. fuscata* (Quiatt, 1979). In both species, a lower-ranking mother was unable to retrieve her infant and it died. Duane Quiatt (1979) says that the kidnapped rhesus infant was "never once mistreated" that he could see, "except by not lactating, an important exception to be sure." As noted by Joan Silk (1980) for captive bonnet macaques, *M. radiata*, kidnapping can be considered a form of female competition, although the lack of lethal kidnappings made that hypothesis debatable (Maestripieri, 1994). Our observations of lethal kidnapping support the hypothesis that the harassment of alien infants by adult females can be a form of reproductive competition between females (Maestripieri, 1994).

The fact that lethal kidnappings occur, however, is not the same as saying that higher-ranking females or matrifocal groups ensure their high rank through this type of "infanticide." Close attention to the circumstances of these lethal kidnappings is essential. Is it the female equivalent of male infanticide? I do not think so. Looking more closely at the lethal kidnappings shows that there is probably no intent to kill. All three of the lethal kidnappers were very protective of the infants over several days. They did not smash their heads on a rock or toss them into the gorge. The kidnappers were not lactating and did not provide food to the newborns. Tubby even took food away from the infant when it got some. Despite what looks obvious to us, that the infant was getting very skinny and starving, it is possible that these monkeys do not perceive that. D. Povinelli, K. Parks, and M. Novak (1992) have shown, for example, in an experiment on role taking, that some rhesus macaques, in contrast to chimpanzees, do not have what is called social attribution, or the ability to perceive what others know. Following another line of reasoning it is quite possible that these babies would have died anyway, even if they were not kidnapped. We have three cases of deaths of newborns because their mothers never lactated. These three mothers were first-time mothers. In one of the lethal cases of kidnapping, the one involving White Eye in Troop Three, the mother was never seen to nurse her baby either. The mother of Tubby's kidnapped infant is unknown. All of the infants in Troop One can be accounted for that summer, and only two other females could possibly have had an infant without our knowing it. Neither of these females had yet had an infant, judging by their nipple length, and could not possibly have nursed an infant.

We did not observe any male infanticide. The killing of infants by males can, in some circumstances, increase the reproductive success of those males in some species (Hrdy, 1979). Some of the circumstances that favor male infanticide can also occur in Balinese and Bornean *M. fascicularis* (Wheatley, 1982). They are rapid male replacement of the alpha rank and a quick return to estrus of the mother whose infant is killed. This species is also rather sexually dimorphic, with adult females approximately 63 percent of the weight of adult males (Schultz, 1956; Washburn, 1942). In addition, the high population density of the Monkey Forest might also be a factor favorable to infanticide. A possible explanation for the lack or rarity of infanticide might be due to a number of things. First, the three troops in the Monkey Forest are multimale. They form coalitions with each other and chase away nontroop males. Second, all of the males in the troop as well as nontroop males are copulating with estrous females. In Troop One, in 1991, for example, nontroop males copulated more, 51.4 percent, with the estrous females in the troop than did the three adult males in the troop. If immigrant males were to kill babies, it would be possible that they, having fathered the infants, would thereby decrease their own reproductive success. Or perhaps the females, by copulating with nontroop males, may be ensuring the absence of infanticide. There were four infant deaths in Troop One that year. Eyelid's young brown male infant, for example, died from massive skull injuries, perhaps suffered when ventral on the mother, who may have hit a branch or the ground in an attempt to flee. The mother Eyelid was seen carrying her dead infant with a severe gash on her arm. Two of the other deaths were of first-time mothers who never nursed their newborns. The fact that deaths of newborns occur during aggressive periods of time when mating occurs does not, of course, mean that males are killing them. Third, the females yield considerable power, especially by giving scream-threats. An infant in distress will elicit mass mobilization of both males and females via scream-threats, and the source of the infant's distress might be instantly attacked. Such coalitions are very important in the maintenance of intra- and intertroop security. High-ranking females as well as low-ranking females come to each other's aid. The sight of Tubby, the 1992 alpha female of Troop One, charging with all her hair out after someone is quite awesome. In 1990, the beta female slapped Two Fingers in the face after he contact aggressed an infant; then she and two other high-ranking females groomed him.

Despite the despotic hierarchies and evident female competition in these monkeys, they are highly cohesive and social. The females grow up and play with each other, and they groom and huddle with each other extensively. Female kin especially groom each other. Everyone, regardless of their rank, grooms mothers with infants. Despite their differences, the females are seen to reconcile and support each other. All of these ties enhance a troop's cohesiveness. While it might appear that appeal aggression such as scream-threats is just another indication of this species' despotism, it is important to emphasize that it is a mechanism of building support and alliances among this female-bonded species. The adaptive value of these alliances (which include males also) that can be instantly mobilized, especially through scream-threats, is obvious. Protection of individuals in the troop, especially the vulnerable infants, is one such advantage. This rapid mobilization force is also used in intergroup competition and not only for purposes of protection. As the section describing intertroop encounters points out, it functions in the defense of and the acquisition of resources.

Dominant Males

Most of the monkeys in the Monkey Forest have physical disfigurements, especially the males. All of the males in the table are recognizable on the basis of serious injuries or disfigurements and were named accordingly. The alpha male of Troop One in 1986, for example, had a middle distal phalange of his right hand sticking up. He also had a broken left canine. The beta male had a crooked tail, reportedly cut by a farmer's sickle. The dominance relationships for males in all three troops over four summers are presented in Tables 3.17, 3.18, and 3.19. The dominance relationships are based on the same criteria as those for the females, that is, threats, grins, and flight. Such agonistic behaviors seem especially common around estrous females.

Table 3.17. **Outcome of Agonistic Interactions among Adult Males in Troop One**[1]

	Subordinate					
Dominant	Alpha	Beta	Two L.E.	Two F.	Cut Tail	Scarface
Alpha 1		28	1	25	—	—
Beta 1,2,3			17	24	1	15
Two Lobe Ear 1					—	—
Two Fingers 1,2,3,4		16	6		16	14
Cut Tail 2,3,4	—		—			3
Scarface[2] 2,3,4	—					

[1] The value in each cell is the frequency that the dominant won an agonistic bout with the subordinate. The numbers after the name in the first row refer to the years that the animal was present: 1 = 1986; 2 = 1990; 3 = 1991; 4 = 1992. — = not present at the same time.

[2] Scarface was the alpha male in Troop Two in 1986.

Table 3.18. **Outcome of Agonistic Interactions among Adult Males in Troop Two**[1]

	Subordinate			
Dominant	Arnold	Tommy	Pointer	Martin
Arnold 2,3,4		5	5	1
Tommy 2,3,4			4	1
Pointer 3,4				1
Martin 4				

[1] See key in Table 3.17. There was only one adult male in the troop in 1986, and by 1990 he had moved to Troop One (Scarface).

Table 3.19. **Outcome of Agonistic Interactions among Males in Troop Three**[1]

	Subordinate		
Dominant	Line Nose	Bent Toe	Curl Ear
Line Nose 2,3,4		17	5
Bent Toe 1,2,3,4		12	
Curl Ear 2,3,4			

[1] For key and explanation refer to Table 3.17.

As can be seen from the tables, there is a linear dominance hierarchy between all males in all troops with one exception involving Two Fingers and the beta male of Troop One. This exception is not really an exception at all but is the result of a dominance reversal between the years 1986 and 1990. In fact, this reversal actually corroborates the rule suggested by Angst (1975) that younger adult males dominate older males. Tables 3.17 and 3.19 also show Scarface and Bent Toe as subordinate males. Both of these older males were alpha males in 1986: Scarface in Troop Two and Bent Toe in Troop Three. Scarface went to Troop One sometime before the summer of 1990 and became the fifth-ranked adult male. His behaviors in Troop One are very interesting as well as unusual, and it may be the case that Troop One was his natal troop. In contrast, Bent Toe remained in Troop Three despite his loss in rank to Line Nose by 1990.

The immigration of Two Fingers into Troop One and his rise to the alpha position is an interesting case to present. Two Fingers was first seen in 1986. As his name suggests, Two Fingers is a very distinctive male. He only has the second and third digits on his left hand. In 1986, his permanent canines had not yet fully erupted. On August 10, for example, his top canines were still growing out and had only just protruded beyond the occlusal surface. I estimate his age at that time to be between four to five years. This estimate is based on the known age of another male born in Troop Two in July of 1986. The Troop Two infant was first seen as a newborn with a herniated navel. Known as Umbilical Boy, his permanent canines had not developed by 1990, but the next year, in 1991, his canines were fully erupted. In 1986, Two Fingers was consistently subordinate to the alpha and beta males, and it was not until 1990 that Two Fingers became the alpha male and was dominant to the beta male. Thus, we have a case in which a male less than nine years of age, several years younger than Angst (1975) saw, came to dominate an older male. The alpha male in 1986 had died by 1990. One of the guards reported that just six months ago, the alpha male, when trying to jump to the ground, had impaled himself on a small tree that had been cut in half.

In 1986 Two Fingers was seen to copulate to ejaculation with almost all of the troop's adult females. This includes copulating with the alpha female and with mothers of unweaned infants. In fact, he was responsible for 18 percent of all the copulations to ejaculation in the troop observed that year. Half of his copulations with the alpha female involved scream-threats from her in which the beta male then assisted her by chasing Two Fingers. In one such coalition with the beta male, the alpha female bit Two Fingers. In the absence of other males, however, Two Fingers copulated with her to ejaculation and was groomed by her afterward on four separate occasions.

Although the alpha and beta males consistently dominated Two Fingers, there were occasions when he did not respond to the beta male's threats. Once both the beta male and Two Fingers inspected the alpha female's genitals at the same time, and Two Fingers then copulated with her several meters from the beta male. The alpha female solicited the beta male in a scream-threat against Two Fingers, but the beta male threatened the alpha female, and then Two Fingers threatened and hit her. She then presented to the beta male, giving a protected threat to Two Fingers, and the beta male walked away. She next proceeded to scream-threat again at Two Fingers, approached and presented to him, and then threatened him once again. When he threatened back, she began her scream-threats at him again while appealing to the beta male for support. In addition, on two occasions in 1986, Two Fingers slapped the beta male while the latter was copulating. While such incidents are not evidence of dominance reversals, because juveniles will also sometimes slap adult males while they are copulating, it is interesting.

Both the beta male and Two Fingers also played together very briefly on several occasions while playing with young juveniles. Once both of these males lipsmacked to each other and grabbed each other's shoulders with an old infant in between them. Afterward Two Fingers held this infant vertically in front of him while the beta male tried to bite Two Fingers. In another case both Two Fingers and the beta male played together and then lipsmacked and grasped each other and "kissed" or put their lips on each other. These and other behaviors are used by males to work out relationships with each other and to indicate their eventual acceptance and mutual support of each other in the troop.

Two Fingers was never seen to hug or grasp the shoulders of the alpha male, but the beta male did so with the alpha male on four separate occasions. Several of these occasions occurred after aggressive interactions. On one such interaction, the alpha male climbed down a tree and grabbed an old infant and pushed it down onto the grass, threatening it. The infant along with four others had just been playing with the beta male. An adult female ran up to the alpha male, giving scream-threats and appealing to the beta male and to Two Fingers less than three meters away. The beta male grabbed the alpha male by the shoulders and lipsmacked rapidly while almost closing his eyes. The alpha male responded with a few slow lip smacks; then the beta male let go of the alpha's shoulders. Next, the beta male grabbed the alpha again, repeated the same behaviors, and withdrew. Two Fingers just looked on. The adult female who had given the scream-threats approached the alpha male and in an apparent reconciliation groomed and grinned at him.

In the other case of mutual shoulder grabbing by males, there was considerable intratroop aggression involving all of the four top-ranking males as well as other males and most of the adult females. The beta male was show looking the alpha male while chasing and threatening the fourth-ranked male, Two Lobe Ear, but eventually the latter male stopped fleeing after the beta's appeal to the alpha male was ignored. The beta male then approached and vigorously grabbed, while standing up, the alpha male around the shoulders with both hands and gave lip smacks. The alpha male also gave lip smacks while grabbing the beta male around the waist, and they then sat down while hugging each other.

Adult males have not always known each other for as long as adult females have known each other because, unlike the females, the males leave their natal troop. The male hugging behavior seems to be a signal of mutual support, and the behavior may be more usual among older males. In addition to the beta male, we also saw Scarface as an older male going through these same behaviors. Scarface and Cut Tail, for example, did a mutual pout face, shoulder grab, and low guttural vocalizations. Coalitions of adult males are important in threatening and chasing strange adult males, as when the beta male and Cut Tail sat, side by side, on a branch, threatening nontroop males such as Umbilical Boy who are copulating with the troop's females. Recall too that the females are scream-threatening these nontroop males, as when Dark Eye female screamed at Umbilical Boy when the beta male discovered them copulating in 1991.

The alpha male had been "boss" since 1978, according to the local people. They also claimed that the beta male came from Sumatra in 1978 with Japanese researchers who worked here. This male was smaller and darker in color than the other males, and he had a black and brownish crest of hair on the back half of his head. Pelage color is indeed generally darker in *M. fascicularis* in Sumatra and Borneo than it is in Java (Fooden, 1995). A local farmer had also cut his tail with a sickle so that it was bent in the middle. This male was the only one that I observed grabbing the testicles of another male. N. Koyama, A. Asnan, and N. Natsir (1981) report this behavior among Sumatran *fascicularis* at Gunung Meru.

Monkeys are probably killed in the fields when they leave the forest. Several farmers told me that they had eaten monkeys. One said that he cooked a monkey with *ikan kambing* (goat fish) and that after eating this dish, his stomach became "so hot"—a sign of spiritual pollution—that he could not sleep. I did see a young man shooting at monkeys at the edge of the Monkey Forest with an air rifle, but I didn't see him hit any, and he was told never to do it again. Monkeys will sometimes come into town, where they are very mischievous. My informants claimed that such animals had been ostracized by their own kind and that they could therefore be killed. A priest had to bless the action, however, before a monkey could be killed with a poisoned sweet potato. Male immigrants from other forests used to come up to Ubud by the riverine gorges, but not recently.

Table 3.17 also shows the adult male Scarface in the Troop One adult male dominance hierarchy. This is another older male who was the alpha male of Troop Two in 1986. He had a scar on the left side of his nose, and his index finger on his right hand was straight. This older male was subordinate to all the males in the hierarchy, although he was seen to chase off Umbilical Boy, a Troop Two subadult male, who was copulating with the females of Troop One. Scarface also engaged in agonistic buffering where, for example, on several occasions he held an infant ventrally and approached the alpha male Two Fingers. In addition, Scarface was seen on the periphery of Troop Three in 1990, threatening subadult males such as Bright Eyes and chasing the troop. Such observations show that, as in other species of macaques, adult males can join other troops during their lifetime.

Male Affiliative Behavior

Adult males typically play with young juveniles. All of the adult males in Troop One, except Two Fingers as an adult, were seen to play. Scarface, the oldest male in the troop, had some of the longest bouts of play. In 1992, he played with four old infants, approximately a year old: the young brown infant of Black Mandible, and one young juvenile for ten minutes. He threatened off the adult female, Point Left Ear, who approached him while he played, and he withdrew from Black Mandible when she also approached. He would gently mouth these youngsters while they played all around and on top of him. Scarface also did the most cases of agonistic buffering, deliberately grabbing youngsters and approaching, while grinning and screaming, at Two Fingers, the alpha male. Scarface is a very interesting and unusual male. He was never seen to copulate with any Troop One female except Lefty in the three years that he was present there. On this one occasion with Lefty, he did not ejaculate. Perhaps Scarface has returned to his natal troop after leaving Troop Two where he was alpha male in 1986.

There appears to be a vocalization that adult males give that elicits approach by juveniles and infants for play. In 1992, the second-ranking Troop One adult male Cut Tail gave what appeared to be a soft kera-like "ah" vocalization similar to a vocalization they do when they see sweet potatoes being offered. After giving this vocalization, a young juvenile female and an older juvenile male immediately approached Cut Tail, and they all played together. Cut Tail playfully bit their feet and their rear ends, sometimes turning them upside down.

There are also observations of what appears to be interference by more dominant males of young immigrant males who initiate play with juveniles and infants of the troop. In 1986 the beta male of Troop One, for example, seemed to follow young Two Fingers around and intervene by supplanting him when Two Fingers tried to play. The beta male followed this up by giving protected threats and chasing Two Fingers. The alpha male then threatened and chased the beta male. Then Two Fingers, followed by the beta male, threatened and chased Two Lobe Ear male who, after being contact aggressed by the beta male, presented and grinned to him. Two Fingers then played with Two Lobe Ear male, and the beta male laid down and played with nine youngsters.

Umbilical Boy is a male in the process of immigration. Born in July of

1986 in Troop Two, he was seen copulating with adult females such as Eyelid in Troop One as early as 1990. He was usually chased away by the adult males of Troop One. Umbilical Boy sometimes associated with and sometimes chased Troop Three from the Garuda area in 1990. In 1991, Umbilical Boy hung around the periphery of Troop One as well as at the Garuda area. He was part of a loosely formed association of subadult and juvenile males that we called the Garuda Boys because they hung out in the Garuda area of the Monkey Forest. The Garuda Boys were aggressive toward tourists, sometimes biting them in unprovoked attacks. In 1986, there were no Garuda Boys, but the rapid growth of the population led to the increasing size of this all-male group. In 1990, there were about twelve subadult and old juveniles, in 1991 about twenty, and in 1992 approximately twenty-two. Many of these Garuda Boys were born in Troop Two.

In 1991, Umbilical Boy was seen copulating to ejaculation with Troop One females again such as Dark Eye and Point Left Ear, and he also groomed the latter female. Umbilical Boy was also seen grooming and being groomed by Cut Tail, the third-ranking male in Troop One. He also continued to chase off Troop Three along with other Garuda Boys and Curl Ear. In 1992, it appeared that Umbilical Boy was starting to get more integrated into Troop One. He was seen subordinate to all the troop's adult males, including the alpha male Two Fingers, the number two male Cut Tail, and the number three male Scarface. By 1992, the old beta male had died and Cut Tail was now second in rank. Umbilical Boy was observed to grin and flee nine times from Cut Tail that year. Umbilical Boy even gave a teeth chatter along with a grin and a lip smack to the approaching hunched stalk walk of Cut Tail. In one interesting interaction, the alpha male Two Fingers threatened a subadult Garuda Boy and show looked to Umbilical Boy. The latter then joined in by threatening that subadult and then Umbilical Boy grinned to Two Fingers. Umbilical Boy continued to copulate with Troop One estrous females such as young number eight female, who also scream-threatened him when Cut Tail was around. These same females also appealed to Umbilical Boy, however, in their scream-threats against other subadult males and occasionally groomed him afterward, as did Young Number Eight female.

So both the males and the females of Troop One were, in 1992, beginning to treat Umbilical Boy as a member of the troop. His integration into the troop was not yet complete because of a number of other observations that especially related to the alpha female Tubby. Some of the females such as Lefty, ranked number ten, still scream-threatened Umbilical Boy and appealed to Tubby against him. Umbilical Boy joined the males of Troop Two in chasing Troop One several times. In another episode, Umbilical Boy helped Troop Two chase Troop One and the Troop One alpha female Tubby gave scream-threats and show-looked to her entire troop. The alpha male, Two Fingers, however, ignored her, and continued to groom White Eye, the number three-ranked female. So Tubby chased Umbilical Boy and some Garuda Boys until they bit her, and she came running back. Toe In, the thirteenth-ranked adult female, and an old juvenile, Hair Face, on two separate occasions during this episode scream-threatened Umbilical Boy and then the entire troop fled from Umbilical Boy. This entire episode finally ended when Cut Tail did an appeal to his troop, giving pout faces and wahoo vocalizations, upon which all the troop's males chased Umbilical Boy. Slightly later, he grinned to Cut Tail, especially after being scream-threatened by the adult female Lefty and stalked by Cut Tail.

Sexual Behavior

Male dominance rank appears to correlate with mating frequency rank. The highest-ranking male, for example, appears to mate the most often. Of course, mating frequency does not necessarily mean actual reproductive success. *Macaca fascicularis* males are interesting in that they can ejaculate either after a single mount or after multiple mounts. Such patterns are important to investigate because they are important in different types of mating systems, and they may reflect phylogenetic relationships. The definition of an ejaculatory sequence is one in which the sequence begins after an ejaculation or after twenty minutes of observation in which the male does not interact sexually with any female (Shively, Clarke, King, & Mitchell, 1982). I only included mounting, thrusting, and intromission (not consorting) as part of the ejaculating sequence, and only copulations in which ejaculation or a distinct pause occurred are considered. All of the reproductive data from 1986 were analyzed. Most of these data were from Troop One males, but some data from the alpha males of the two other troops were also included. All of Troop One's adult and subadult males and females were seen to copulate.

The 1986 data show that 82 percent of the total number of mounts-to-ejaculation ($N = 169$) was of the single mount-to-ejaculation type, and 18 percent was of the multiple mount-to-ejaculation type. In Troop One, the alpha male did 50.9 percent of all copulations to ejaculation, the beta male did 31.1 percent, Two Fingers did 18.0 percent, and another younger male did .01 percent. Table 3.20 shows that the mean time of copulation-to-ejaculation was 7.6 seconds for both adult males, whereas the mean for Two Fingers was 8.1 seconds. The alpha male did 55.2 percent of the total multiple mount-to-ejaculation sequences, the beta male did 34.5 percent, and the old juvenile male did 10.3 percent. The range for the duration of the multiple mount-to-ejaculation sequence was 2 to 35 minutes, with a mean of 7.1 minutes. The range of the number of mounts for the multiple mount-to-ejaculation sequence was two to six mounts, with a mean of 2.37. Ninety-seven percent, or 126 of all copulations to ejaculations, had a female copulatory call, whereas only 3 percent, or four cases, had no call.

The data on the copulations in Troop One for 1992 show that Two Fingers, the alpha male, did 55 percent of all copulating to ejaculation, while Cut Tail male (the second-ranked male) did 31 percent ($N = 49$). About five other subadults and adults make up the remainder of the cases. All but three of the cases were of the single mount to ejaculation type (94 percent). Two Fingers

Table 3.20. **Copulation Sequence Data Collected in 1986 on Males in the Monkey Forest**

	SME: Range	Mean	MME: Range	Mean
Duration of sequence	2.6–12.0 sec.	7.1 sec.	2–35 min.	7.1 min.
Duration of mount	2.6–12.0 sec.	7.1 sec.	3–9.4 sec.	6.425 sec.
Number of mounts	1	1	2–6	2.37
Number of thrusts/mount	5-32	16.36	6–31	15
Thrusting rate/mount	1.0–4.23	2.37	.94–3.33	2.34
Number of thrusts/sequence	–		16–67	33.65
Number of males		5		4

SME = single mount to ejaculation ; MME = multiple mount to ejaculation.

mounted eight times over a thirty-six minute period until ejaculation, and Cut Tail had the two other cases of multiple mount to ejaculation. He mounted twice over a six-minute period and three times over a twenty-two minute period. Two Fingers ran off chasing other males after three of his eight mounts. It may therefore be the case that a higher-ranking male can monopolize longer, however briefly, an estrous female through multiple mountings than can a lower-ranking male. The problem that a subordinate male has with multiple mounting can also be seen in an example in 1991. Bent Toe, the beta male of Troop Three, copulated with Scraggle Left Ear, a mother. There were fifteen thrusts in 3.4 seconds, no ejaculation, and no call. She moved off, and Line Nose, the alpha male, approached her and copulated with her without ejaculating six minutes later.

There appeared to be sexual partner preferences. Two Fingers copulated mostly with Tubby, the alpha female, and with Toe In, a very low-ranking female. The alpha male's copulations with Tubby, the alpha female of Troop One in 1992, is a dramatic change from the previous two years, when neither he nor anyone else was ever seen to copulate with her. In the previous years, Tubby was not alpha female, and she is believed to be sterile. The former alpha female died during childbirth before our 1992 fieldwork, according to the guard. Cut Tail copulated mostly with Point Left Ear and Young Fifteen Female and only once with Tubby and Toe In. Cut Tail averaged 17.9 thrusts per mount to ejaculation (N = 9). Point Left Ear copulated with many different adult and subadult males.

The following data were collected in 1991 on Troops One and Three. Troop Three, with its three adult males, had 69.6 percent of its total copulations as single mount to ejaculation and 30.4 percent as multiple mount to ejaculation (N = 23). The alpha male, Line Nose, did 52.2 percent of all copulations to ejaculation followed by the beta male, Bent Toe, with 13 percent, and finally Curl Ear with 34.8 percent. Line Nose did five out of the seven multiple mounts to ejaculation seen in this troop (71.4 percent). Most of these were longer in duration, as long as thirty-two minutes, and had more mounts (seven). Line Noses's average time to ejaculation during the single mount pattern was 7.1 seconds ($N = 4$) which was longer than the older beta male, Bent Toe (6.65 seconds, $N = 2$) and Curl Ear (6.45 seconds, $N = 6$). There were a few patterns to partner preferences, with all eight of Curl Ear's ejaculations with Bend Tail, the lowest-ranking female of the troop, and the alpha male with all of the estrous females, especially Scraggle Left Ear.

In 1991 for Troop One, the reproductive data show many more copulating males than were in the troop. The alpha male, Two Fingers, did the most copulations in the troop, with 28.6 percent, the beta male did 9.5 percent, and Cut Tail, the number three male, did 9.5 percent. A nontroop male named Black Balls did 24 percent of all ejaculations. Umbilical Boy did 7 percent, as did another subadult male, and the remaining copulations to ejaculation were by various males, the so-called Garuda Boys, in the Garuda area. Most of the forty-two ejaculations in the troop were of the single mount pattern (88.1 percent); the rest were multiple mounts to ejaculation. There were a few partner preferences. The alpha male preferred the alpha female, followed by Toe In, but the alpha female was copulated with by six different males, especially Black Balls. Notably absent in all years of data collection with the exception of 1992 were any copulations with the sterile Tubby. The longest multiple mount sequence was by Cut Tail with White Eye, the second-ranking female. It lasted thirty-nine minutes and included eight mounts.

In Troop Three in 1990, the alpha male, Line Nose, made 78 percent of all copulations to ejaculation, with three other males doing the remainder ($N = 25$). Sixty-four percent of all copulations were single mount to ejaculation, and Line Nose did 78 percent of those. The average duration of all copulations was 7.69 seconds ($N = 14$). For Troop One, the alpha male, Two Fingers, did 81 percent of all multiple mounts to ejaculation. Most of the copulations in the troop were of the single mount variety (72.2 percent). The average duration of copulations was 5.9 seconds ($N = 18$). Both alpha males in the two troops copulated with all the estrous females in the troop.

Summary of Male Rank and Sexual Behavior

High rank, especially the alpha rank, enables males to have sexual access to estrous females. The alpha males did 54 percent of all copulations to ejaculation in Troops One and Three over the four summers of research ($N = 342$). Each of these troops was multimale, with at least three copulating males, and the mean was five copulating males per troop. These observations and others are very similar to wild *M. fascicularis* (Wheatley, 1982). Many of the dominance interactions, such as the grinning and chasing between males, took place near estrous females. Estrous females advertise their sexual receptivity through their copulatory calls. Females gave copulatory calls 93 percent of the time ($N = 254$). Such calls can be given between mounts of multiple-mount-to-ejaculation sequences, so that the calls do not always imply ejaculation. On two occasions, we saw a young juvenile female who did not copulate give a call next to an adult male, who responded by threatening her.

Estrous females often copulate with many different males. For example, in 1991, the Troop One alpha female copulated to ejaculation with at least seven different males, including all the adult males in the troop with the exception of Scarface, the older male. She also copulated to ejaculation with many nontroop males. Sometimes she mated with different males within a nineteen-minute period of time. Mating with only one male is probably exceptional in these Monkey Forest females.

This species of macaque is interesting in that males sometimes engage in single mount to ejaculation and sometimes in multiple mount to ejaculation. The former was more typical, with 81.3 percent of all copulations, whereas the multiple-mount pattern comprised 18.7 percent ($N = 342$). Alpha males also did 60.3 percent of all of the multiple mounts ($N = 63$). The multiple mounting as well as the grooming of the female afterward not only function in monopolizing a female but also to prevent loss of sperm and perhaps facilitate the formation of a plug. Interestingly, in wild *M. fascicularis*, the female invariably groomed the male after copulation (Wheatley, 1982). It is difficult for a low-ranking male to monopolize an estrous female for any length of time around higher-ranking males. Estrous females, with their copulatory calls and scream-threats, also add to this difficulty. We have seen higher-ranking males, who proceeded to copulate to ejaculation with that female, displace lower ranking males in the midst of multiple mounting. The longest multiple mount-to-ejaculation sequence ever recorded was in 1990, when the alpha male of Troop One, Two Fingers, mounted White Eye thirteen times over a fifty-six minute period.

Another very interesting observation is that an adult female of Troop One named Tubby was *never* seen to copulate in 1986, 1990, or in 1991. Tubby, however, became alpha female in 1992, and in that year, one-third of the entire alpha male's copulations that we observed was with Tubby ($N = 27$). Cut Tail was the only other male to ever copulate with her and this occurred only one time. Tubby never had a nursing infant, although she did kidnap them. The role of the alpha female is probably very important in the power structure of the troop for the alpha male to mate with her rather than with other available estrous females.

Birth Season

Births are nonseasonal in the Monkey Forest. During the four summers of research, we have seen fourteen newborns (black infants) in the three troops. By looking at the ages of infants when they were being weaned and the numbers of estrous females in the summer, we can reconstruct an approximate time of birth. Table 3.21 shows the reproductive condition for the females of Troops One and Three for four summers (1986, 1990–92) and for the females of Troop Two for three summers (1990–92). About half of the females in these troops are in estrus during the summer, and the other half is still nursing older infants. It appears that births are somewhat bimodal, with peaks in summer and in winter. This coincides with the two tourist seasons—summer for North Americans and Europeans and winter for Australians. Births probably occur in all months of the year. A female can return to estrus very quickly, sometimes within a month after the death of her baby, as was the case with Dark Eye (an older female present in 1986) after her newborn male died in 1991.

Three births have been seen, two on the ground and one in a tree. The mothers licked the baby and ate the placenta. The babies seemed very alert, opening their eyes and clinging within half an hour of birth. Miss Digit, ranked the second to the bottom in Troop Three in 1991, gave birth on July 26, 1991, while standing on a branch at the top of a tall fig tree. She screamed during the birth, and some animals, including Bend Tail, the bottom-ranked female, attacked a nearby subadult male.

Table 3.21. **Cycle Status of Adult Females in Troops One, Two and Three as Observed during the Summers of 1986, 1990, 1991 and 1992**[1]

Year	**1986**		**1990**			**1991**			**1992**		
Troop	**1**	**3**	**1**	**2**	**3**	**1**	**2**	**3**	**1**	**2**	**3**
***N* of Adult Females**	**9**	**7**	**12**	**11**	**6**	**12**	**14**	**9**	**14**	**17**	**11**
Estrous—cop./ejac. with adult males	2	0–2	5	3	2	5[2]	5	3	5	9	4
Newborn (black infant) birth during obs. period	1	1	0	1	2	3	3	1	1	1	1
Brown infant, nursing	5	4	5	6	1	3	5	4	7	4	5
No estrus, no infant, possibly pregnant	0	0–2	1	0	1	1	0	1	0	2	1
Presumed sterile, inf. 1 never obs., short nipples	1	0	1	1	0	1	1	0	1[3]	1	0

[1] Insufficient 1986 data for Troop Two to add to this table.

[2] The new alpha female, Tubby, in Troop One was observed copulating with the troop males, including the alpha male, during the summer of 1992. She was not observed in any copulation prior to this year when she became alpha upon the death of the former alpha.

[3] Dark Eye came into estrus after the death of her newborn.

Vocalizations

An adequate understanding of a species' social behavior and organization is not complete without integrating its vocal communication. Although the vocal repertoire of macaques has been investigated more than that of any other group of primates, it is still inadequate, especially for some species such as *M. fascicularis*. The vocal repertoire of a species needs to be built from different populations as well as from diverse habitats. Only in this way can we understand how different selective pressures have influenced the repertoire. Ideally, this information requires an intimate knowledge of a social group, such as the identity of not only the emitter but also of the receiver, and other aspects of the social situation. Such knowledge can usually only be obtained with habituated animals and good visual and acoustic conditions, often at close range. Such conditions are easily met in the Monkey Forests of Bali, where tourists and local people, as pointed out previously, have interacted with the monkeys for generations if not centuries. The only previous field study of the vocal communication of this species emphasized the louder calls (Palombit, 1992a; 1992b). Our research on vocalizations complements previous work and expands our understanding of the species by examining calling behavior in the field, in social situations, at close range and in an environment where troops frequently interact with each other as well as humans.

Vocalizations were sampled *ad libitum* during the 1992 field session in June and July. Three different observers recorded animals in all three troops. Recordings were made within a few meters of the calling animal, using small cassette tape recorders. The identity of the caller and the potential receiver(s) and the context of the call were also recorded. The identity of the receiver of the call was usually obvious based on the receiver's proximity and/or response to the call. A Kay Digital Sonagraph 7800 with a wide band filter was used to analyze the acoustic structure of the vocalizations.

The following data extend our knowledge in several ways. First, we describe ten new calls for this species. The larger sample sizes and the recording of vocalizations close to the source of interacting individuals pick up many new and different types of calls other than just the more conspicuous and loud calls. Many of these calls have been described before in other species of macaques. These are: the threat rattle and pulsed whoos in *M. radiata* (Hohmann, 1989); the grunts of *M. sinica* (Dittus, 1988); and the various screams, such as the noisy, undulating, tonal, and pulsed screams of *M. mulatta* and *M. nemestrina* (Gouzoules & Gouzoules, 1989; Gouzoules, Gouzoules, & Marler, 1984). Second, we describe a new scream that we call the banded scream, named after its structural appearance (see Figure A3b). Third, our vocalizations were recorded several meters away from the sender. Such conditions are likely to pick up the high frequencies, which tend to degrade under other conditions (Wiley & Richards, 1978). For example, Marc D. Hauser (1993) evaluated the effect of phylogeny, body weight, and social context on vocalizations using Palombit's (1992a) data on *M. fascicularis* without such calls with higher frequencies such as tonal and noisy screams. Unfortunately, these calls were not described by Palombit, so we do not know if our data represent habitat differences, population differences, or observation conditions. Fourth, we describe some of the contexts in which these calls are given. Such information is often difficult to differentiate, given the many different nearby individuals of different ages, sex, and rank. Playback experiments will need to be done to better understand the type of specific information that these calls can provide. The different screams are important in the recruitment of allies, as Gouzoules and

Gouzoules (1989) found. Given the cognitive development of these primates, the many variants and intergradations between the calls stereotyped by our approach, and that of others, imply a richness of meanings (Green, 1981; Peters, 1986). Last, our data add to, and in some cases fail to corroborate, previous work on the species (Palombit, 1992a; 1992b). For example, the wahoo loud call is a very distinct call. It is, however, very similar to the wahoo, or type 2, loud call found in baboons, according to the description and spectrogram by Richard W. Byrne (1981). The wahoo in *M. fascicularis* is, therefore, neither unique to the species, nor is it among the longest in duration, as previous work suggested (Palombit, 1992a). The wahoo appears to regulate intragroup spacing by bringing individuals together in the context of intergroup disturbances (pers. obs.). Many different types of calls that we recorded averaged longer durations, such as tonal, banded, and undulating screams, grunt-coo, and segmented coo. Our longest call was a copulatory call, which was 8.2 seconds long.

Sixteen different vocalizations were classified into nine categories based on previous descriptions of the calls of the genus *Macaca*. Spectrographic analysis of 750 calls was made, of which 616 were utilized below. Vocalizations are presented in more detail in the appendix. We were unable to obtain recordings of all vocalizations. For example, one was a particular call used by adult males to solicit play from infants and juveniles. Another was a mmrrh, or trill vocalization used by several adult females, including the mother, on two occasions when a young infant went down into an irrigation canal. Upon hearing the females' vocalization, the infant climbed back out, and the mother approached and retrieved it. The vocal repertoire includes:

1) *Kra call.* This is the call from which the species receives its common name. These calls occur in alarm situations. The various characteristics used by Palombit (1992a) in defining the different types of kra calls were analyzed. Our large sample ($N = 109$) of this call shows that most of our calls have overlap in the variables used by Palombit (1992a) in differentiating kra-c calls from kra-a calls in Bornean *Macaca fascicularis*. See Figures A1a and A1c and Table A1 in the appendix.

2) *Threats.* These are calls that occur in aggressive interactions. After examining the sonograms, we delineated three types:
   a) Threat rattle. These calls are similar to those described by Hohmann (1989) for *M. radiata*. These calls are rapidly pulsed ($N = 29$). See Figure A1e in the appendix.
   b) Bark kra. This is another temporally segmented, or pulsed call as described by Palombit (1992a) (N = 5). This is a high-intensity threat during aid giving, accompanying such events as screaming and show-looking. See Figure A1d in the appendix.
   c) Bark. The pulses are fused in this call ($N = 4$). See Figure A1b in the appendix.

3) *Krahoo.* This is a low, loud, continuous two-syllable vocalization usually given by adult males. It appears to be similar to baboons (Byrne, 1981;

Kudo, 1987). It is a rapid and forceful exhalation followed by a rapid and forceful inhalation. The context of twenty-seven of these calls was obtained during the summer of 1986. They are given during disturbances, such as when gamelan is heard, when dogs of farmers are sighted, and during intratroop disturbances (*N* = 16). Animals rushed to the caller after the call was given (pers. obs.). Ninety percent of the calls were given by the alpha male (*N* = 19). Only once was an adult female seen to give the call. In 1992, only adult males gave the call; alpha males gave eight and a beta male gave eleven. These calls were sometimes accompanied by other screams (*N* = 10). See Table A2 and Figure A2b in the appendix.

4) *Screams.* These were continual vocalizations as described by S. Gouzoules, H. Gouzoules, and P. Marler (1984) for *M. mulatta* and Gouzoules & Gouzoules (1989) for *M. nemestrina*. There are five kinds of screams:
   a) Noisy scream. This is an atonal noisy call given by "victims" of contact aggression (*N* = 26). See Figure A2d in the appendix.
   b) Khreet screech. This is a call of sharply ascending and descending frequency modulations of one or more peaks, and it is similar to Palombit's (1992a) call (*N* = 68). See Figure A2a in the appendix.
   c) Tonal scream. This is an ascending and descending tonal call (*N* = 14). See Figure A2e in the appendix.
   d) Pulsed scream. This is a pulsed call of short duration (*N* = 9). Most of these calls seemed directed toward larger individuals, as if the caller was soliciting aid. See Figure A3a in the appendix.
   e) Banded scream. This long call is a previously undescribed scream. Its name derives from its structure as it appears on the spectrogram (*N* = 83). See Figure A3b in the appendix.

5) *Copulation call.* This is a staccato call given by females during copulation (*N* = 22). Such calls can be the longest of any call in the repertoire. One of our calls was 8.2 seconds long. See Figure A3c in the appendix.

6) *Affiliation call.* This call is structurally similar to the call described by N. Masataka and B. Thierry (1993) for *M. tonkeana* (*N* = 10). Females gave these calls when attempting to handle or get close to infants. See Figure A4a in the appendix.

7) *Contact or coo call.* We found two basic types of coo calls:
   a) Coo. Sonograms revealed many different types of coos. These calls seem to promote friendly interactions and avoid aggression as previously described by Steven Green (1975) and Ryne Palombit (1992a). Most of our coos had no modulations unlike Palombit's (1992a) description (*N* = 18). Subordinate adult females gave this call to dominants; adults gave when isolated from others, and infants called to their mothers. Other coos had more harmonics and the various contexts included an adult female grooming a dead infant, a juvenile trying

to nurse and infants separated from their mothers ($N = 6$). Other coos had segmented bands, called pulsed whoo calls by Hohmann (1989) for *M. radiata* ($N = 4$). This call occurred when a more dominant adult female held an infant of a low-ranking mother. See Figure A4c in the appendix.

b) Wraagh or grunt-coo. This coo had pulses similar to grunts ($N = 65$). These calls occurred during troop movement and during feeding contexts, such as when tourists were feeding bananas to them. See Figure A4d of the appendix.

8) *Grunts*. This call has a series of rapid staccato pulses. It is very similar to the grunts of *M. sinica* (Dittus, 1988), *M. arctoides* (Bertrand, 1969) and *Cercopithecus aethiops* (Cheney & Seyfarth, 1990). See Figure A4b in the appendix.

9) *Geckers*. These calls are short and plosive. Infants give these calls during weaning conflicts with the mother, or in other situations of conflict. For example, an old infant gave this call when it was not allowed to nurse.

Different environments pose different problems for vocal communication. Troops of individuals living in large home ranges in which they are commonly dispersed into subgroups face different problems than troops of individuals living in small home ranges in which the population is highly dense and in constant intra- and intertroop visual contact. The greater differentiation of clear calls and vocalizations associated with foraging, when such stereotyped calls may reflect their more basic functions in protection from predators and food acquisition may be related to a dense, rain forest habitat. Masataka and Thierry (1993) say, for example, that the more frequency-modulated calls occur as distances increase between the callers and the other troop members. Likewise, it may be the case that such calls as screams are more differentiated in monkey forest habitats or where group size increases (Van Schaik, Van Noordwijk, Warsono, & Sutriono, 1983). Palombit (1992b) did not find positive correlations of agonistic vocalizations such as "the scream" or the "Khreeet screech" and party size. The closer proximity of different matrifocal groups in the monkey forest may contrast to the dispersed matrifocal subgroups of the rain forest.

Intertroop Behavior

Two remarkable events occurred during the second month of my 1986 observations. These events are best described as battles between two troops. They were brief, but they involved almost all of the troop members, including mothers of newborns. These two encounters were between Troops One and Two. Usually Troop One supplanted Troop Two during intertroop encounters, but on two violent occasions, Troop Two fought back. An aggressive intertroop encounter is when agonistic behavior (chase, flight, and so on) occurs between members of different troops. The first battle occurred in the Temple Forest

Area of the Monkey Forest on July 14, and the second one occurred on July 29. I will present details on the second encounter because it occurred in the Garuda area, where visibility was good.

Intertroop Encounters

The encounter began at 9:45 A.M., when Troop One started moving up the Gate Road from the Garuda Area to the Feeding Area. Troop Two had finished feeding there and was moving down toward the Garuda Area. Eleven adults in each troop formed a line, shoulder-to-shoulder, facing each other. The beta male of Troop One charged toward Troop Two, and the Troop One alpha male threatened an estrous female. Initially, Scarface, the Troop Two alpha male, was sitting on a log, but he quickly threatened both that estrous female and another one in his own troop and then charged and lunged at the alpha male of Troop One, but no contact aggression was made. Four Troop Two adult females with old infants hanging on then chased Troop One. Two mothers with newborns hanging on ventrally were also present in this encounter, and the beta male slapped one of these mothers twice. The Troop One alpha and beta males and two females charged Scarface, and he withdrew to the stairs going down to the gorge and gave wahoo vocalizations. The frequency of slaps (contact aggression) totaled thirteen in the first four minutes of the encounter. Scarface then returned to chase the alpha male, and a Troop Two mother with an infant also threatened the alpha male. At 9:52, a young juvenile male with only one arm from Troop One was upended and immediately mobbed and bitten by four or five individuals from Troop Two, none of whom were adult males. The same happened to another Troop One individual. A Troop Two mother with her newborn slapped someone and then she and another mother also with a newborn slapped the beta male. The beta male slapped back and advanced, along with the alpha male. Three Troop Two adult females, however, scream-threatened the beta male and stopped his advance. At 9:55, Scarface copulated with one of his Troop Two females, and the alpha male from Troop One also copulated with one of his troop females, probably with the alpha female. Several Balinese boys used their slingshots to try to stop the aggression. The Troop One beta male, however, ran over and slapped a Troop Two adult female and bit another. Scarface copulated again. The beta male from Troop One charged again, and Scarface retreated down the road. The beta male then charged two more mothers with their infants and Toe In (the lowest-ranking adult female in Troop One) also threatened one. At 10:02, the alpha male went to the front of his troop, close to Troop Two, but Scarface and his troop retreated. During the twenty-minute encounter, there were twenty-six occurrences of contact aggression, a rate of 97.5 per hour.

The hour after this encounter was also interesting as the members of Troop One rested on the Temple Forest Hill, all within five meters of each other. At 10:04, the beta male was seen to slap Two Fingers (a juvenile at that time), and then they hugged. Three adult females, including the alpha female, groomed the alpha male. The beta male was groomed by Two-Lobe Ear (another male). The rest of the animals were huddled up, grooming each other. Toe In (the lowest-ranking female) was licking the open wounds of an adult female. The alpha female also groomed the one-armed juvenile, who was seriously wounded, with blood on his back, his left side, and his tail. I was very surprised to see that he survived. One notable behavior is the extensive licking of wounds by other animals.

The other encounter between Troop One and Two occurred on July 14, 1986. Troop One was foraging on grass near the first weir and toward the Pura Dalem. At 2:48 scream-threats broke out near the Pura Dalem, and most of the troop immediately rushed 40 meters on the ground to the Holy Beringin near the Pura Dalem in the Temple Forest. Two phalanxes of individuals formed in lines, shoulder-to-shoulder, and faced each other, eleven adults from Troop Two and ten from Troop One. Troop Two was advancing, pushing Troop One back. Troop Two mothers with newborns were shoulder-to-shoulder, charging and slapping at Troop One individuals. The alpha male of Troop One seemed to charge back and forth with Scarface, the alpha male of Troop Two. Only one large adult female of Troop One did not get involved and sat by a tree watching all the action. At 3:00, Scarface gave wahoos about ten meters away, and Troop One quickly advanced; two adult females of Troop Two then charged the alpha male, who retreated three meters. The beta male next charged and then retreated three meters. Scarface ran to the front of his troop and both troops faced off again, one meter apart, charging back and forth while giving scream-threats. Two adult females faced the alpha and beta males, who were within a meter, and the alpha male retreated. At 3:03, an alarm call was given, and most of Troop One climbed into the trees. The alpha and beta males alternated charging a meter and stopping. At 3:04, Troop Two mothers with their newborns ran up to the front line, and both the alpha and beta males of Troop One withdrew four meters. Three Troop Two adult females together charged three adult females from Troop One, and the latter troop retreated a total distance of about fifty meters, down into the middle of a gully. Troop One then advanced back up the hill out of the gully. At 3:05, the alpha male went up a tree, and the beta male stayed at the bottom. Scarface also went up a tree, and four adult females joined and groomed him. The others in both troops all climbed into trees, and the encounter was over. More than fifty animals had participated, giving scream-threats and screeches for a total of nineteen minutes. Three animals from Troop One were injured during this encounter. The beta male and a juvenile male had wounded tails. An adult female had a cut chest and tail and was missing some hair on her thigh. Animals in Troop Two were also wounded. An hour later, Troop One went to the Feeding Station and ate sweet potatoes. Troop Two moved to the Garuda area and started to go to the feeder but stopped. At 4:26, the alpha male and the rest of Troop One went up a small cliff by the feeder as if avoiding Scarface's troop.

These two intertroop encounters are very important observations in a number of ways. First, intertroop encounters may be somewhat infrequent and rarely observed. Second, they are often chaotic and occur where observation conditions and habituation are not good, and it is difficult to know what is going on. Third, data are rare on the role of females, and their involvement in intergroup aggression has been questioned. As just described, adult females can be extensively involved in these encounters. Not only single adult females, but also mothers with their newborns were seen to charge adult males of the other troop and to slap them! Much of the dearth of information on females is because it is difficult to obtain. Females are usually more difficult to observe and to identify than males. Males are more visible, make a lot more noise and are far fewer in number than females. The intense drama of intrasexual competition between males is distracting.

Such observations warranted further work on these monkeys, and I resolved to return to take a closer look at intertroop involvement. How important are these encounters? Are females defending or acquiring resources? Why might they risk injury to their newborns and older infants by attacking adult males?

Troop Sizes and Core Ranges in the Monkey Forest

According to the local villagers, monkeys have always been here and associated with the main temple or Pura Dalem of this forest. The temple is at least 300 or 400 years old, they say. A number of local villagers maintained, in separate interviews, that the total population of monkeys in 1968 was 23 (plus or minus one), probably all living in one troop. Around the onset of provisioning in 1976, the monkey population was said to be twenty-five. In 1978, N. Koyama, A. Asnan, and N. Natsir (1981) reported thirty-one animals in one troop. There were eleven adult females and one adult male. In 1980, they report that this troop increased to forty-one individuals, with the same one-eyed adult male as alpha. By 1986, when I started my observations, there were sixty-nine animals in three troops. It seems quite likely that Troop One was ancestral in some way to the other two troops. It had the oldest females, and it ranged in the areas where provisioning most occurred. Troop Three was much smaller in size and ranged outside the forest sanctuary more than Troop One. Immigrant males from other areas were said to come into the Monkey Forest, perhaps as late as the 1970s, but this had apparently stopped by the 1980s. Koyama, Asnan, and Natsir (1981) report that two adult males probably immigrated into the Ubud forest from a village about four kilometers from the south. From 1986 to 1992, the population size increased dramatically, up 133 percent in 1992. Table 3.1 also gives the age-sex composition of the three troops for the years 1986, 1990, 1991, and 1992.

The core ranges for the three troops can be constructed from the scan samples taken at half-hour intervals on each troop during the last week of the year's research. In 1990 and 1991, I took the scan samples of Troop One and Troop Three while a research assistant, Katy Gonder, took samples of Troop Two. In 1992, I took samples of Troop One, Katy Gonder of Troop Two, and Tripp Holman of Troop Three. The core range of a troop was defined as an area where the troop frequented at least 50 percent of all known sightings in that area by all the troops. The other two troops would, therefore, together make up less than 50 percent of the scan sightings in that area. In 1990, these areas were broadly defined into about twenty such areas. In 1991 and 1992, however, more precise areas were delineated, using a grid system of 20 meters by 20 meters. There was extensive overlap in home ranges. For each troop, there was probably complete overlap of its range by the other two troops.

The map that follows shows the core ranges for Troops One, Two, and Three in 1990. The 1990 core range for Troop One was between the ranges of the other two troops. Troop One's core area extended from the Gate Road to the Back Road, across the Graveyard and into the Temple Forest. The core range for Troop Two was in the Northwest Gorge: in the forest, Bathing Temple, the Temple Road, the Garuda Forest, and the Pura Dalem. Troop Three's core range was in the Southeast Gorge from the Back Road to the Padmasana and along the weir system on the edges of the rice fields. The 1990 daily ranges of these troops was very short, averaging 450 meters in all. The troop ranges were probably about the same for the year of 1986, based on their sightings for that year. The scan samples also showed that Troop One's daily range averaged 650 meters over seven days, Troop Two averaged 400 meters over ten days, and Troop Three averaged 350 meters over nine days. An interesting difference between Troop One and Troop Three showed up on the proportion of arboreality. On average, over 50 percent of Troop Three were arboreal, compared to 25 percent for Troop One. Troop Three appeared to use trees, especially in the ravines, as areas of safety.

In 1991 an interesting development for Troop Three took place. It was not just confined to the one gorge in the southeast, but had spread to the Northwest Gorge as well. By 1992, Troop Three's range would only be in the Northwest Gorge, as Troop Two appeared to have spread further east. Troop One's core range in 1991 went from the Garuda Area through the Graveyard and beyond the Back Road. Troop Two's range appeared to be expanding a bit into the Temple Forest and across the Gate Road. Troop Three was the most compact troop. Its diameter was forty-one meters across, whereas the other two troops averaged fifty-two meters. Troop Three also had the fewest number of terrestrial individuals. Troop One had the most terrestrial individuals and the fewest arboreal individuals.

The core ranges in 1992 were very different from 1990 and 1991.Troop Two was the most centrally located troop, rather than Troop One, which was that in 1990. Also in 1992, Troop Two ranged from the Gate Road across the Graveyard and into the Garuda Area and Temple Forest. Troop One was ranging along the Back Road, and Troop Three was now centered on the Temple Road in front of the Pura Dalem.

Intertroop Dominance. The three troops in the Monkey Forest had a consistent dominance relationship up until 1992, when a reversal in dominance was apparent. During the four research sessions, every intertroop dominance interaction was noted. In order to facilitate comparisons on rates of encounters between years, I will include only those that took place from my notes. The occurrence of an aggressive troop-troop encounter was scored as

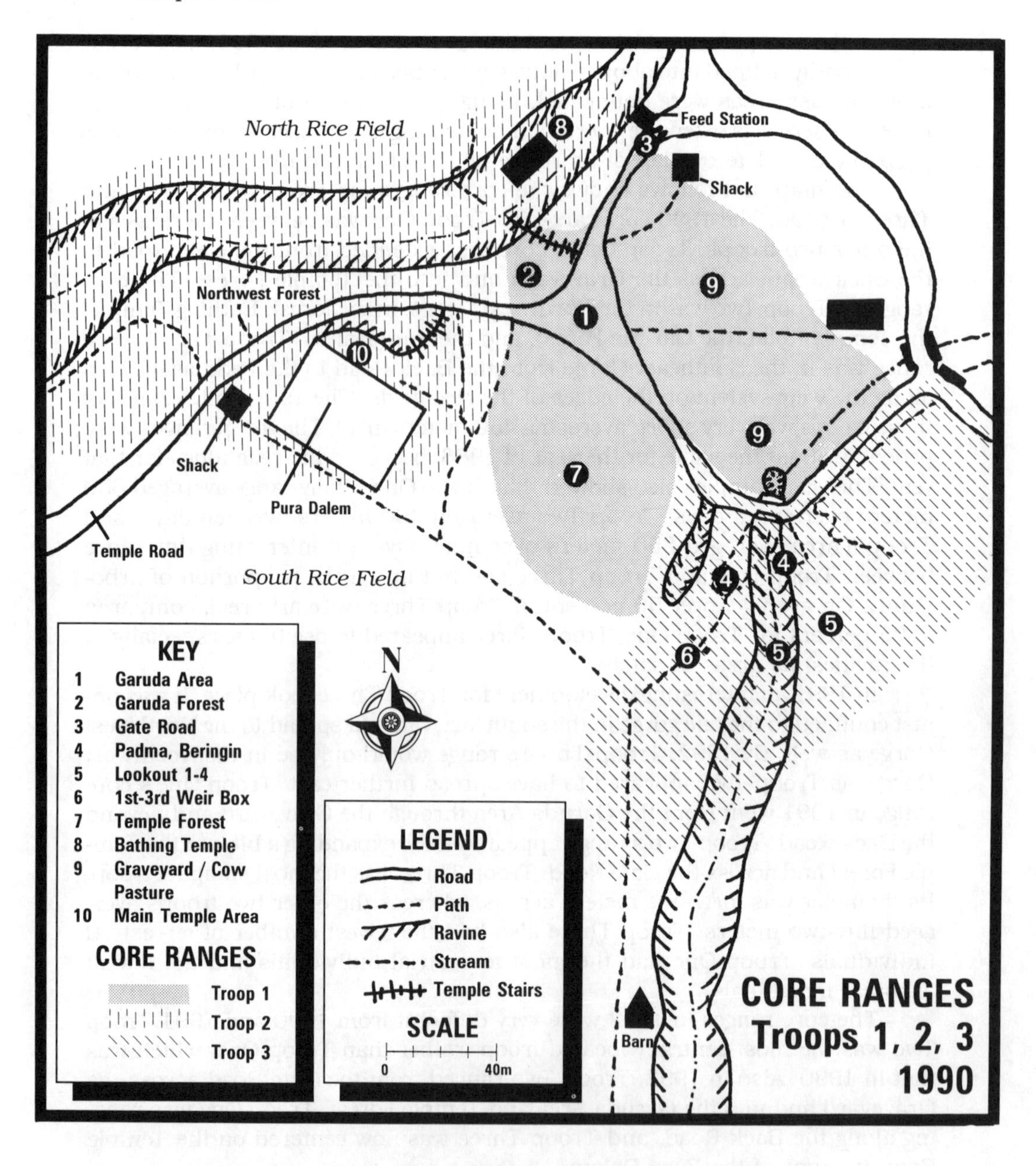

one where aggression took place between members of different troops, whereby the aggressors supplanted the losers.

In 1986, I saw twenty-two cases of aggressive intertroop encounters. This is a rate of about 1/15.5 hours of observation, or about one every three days. The troop-troop dominance hierarchy was linear and stable; that is, Troop One dominated Troop Two and Troop Two dominated Troop Three. There was one encounter between Troops One and Two that was indecisive. This encounter was very violent, and it occurred on July 14, 1986 (see the previous section for a description of this encounter). Although Troop One initially gave up ground during the encounter, by the end of it Troop One had regained its previously "lost" ground. That summer was the last time that such violent intertroop encounters were ever seen. We saw animals get bitten by other troop individuals during encounters, but not nearly at the same level of aggression as the two incidents in 1986.

In 1990, I saw thirty-five cases of aggressive intertroop encounters. Troop One supplanted Troop Two eight times, Troop Two supplanted Troop Three once, and Troop One supplanted Troop Three on twenty-six occasions. The circumstances of these encounters are interesting. Adult males by themselves or coalitions of adult males were seen to effectively chase off troops. Both Two Fingers, the alpha male, and Old Crooked Tail, the beta male, of Troop One, for example, chased off Troop Two twice by themselves, but in these cases, it was the Troop One females who initiated the encounters by scream-threatening. Several times, Two Fingers ran over when his troop was scream-threatened by adult females of Troop Two. Another interesting set of observations was on an adult female of Troop One named Lefty, who appeared to go out of her way to harass and chase Troop Three. Once she gave scream-threats and was backed up by the beta male, who then chased Troop Three. On another occasion, she chased Troop Three by herself. Lefty is handicapped, having only one arm, and she was also somewhat low-ranking at number eight out of twelve adult females in Troop One in 1991.

In 1991, there were twenty-two aggressive encounters with troop-troop supplanting. This is a rate of one for every 6.15 hours of observation. In 1992, there were twenty-one aggressive intertroop encounters, a rate of one for every 6.7 hours of observation. By 1992, there was a dominance reversal between Troop One and Troop Two. Of the twenty-one aggressive encounters that I witnessed, eighteen were of Troop Two supplanting Troop One. Troop One also fled twice from the all-male Garuda Boys. Only once was Troop One seen to displace Troop Three, but these two troops now rarely met because Troop Two ranged in between them. The two research assistants that year also saw an additional thirteen encounters. Troop Three fled from and was displaced by Troop Two on ten occasions, and there were three more cases of Troop Two supplanting Troop One.

Fieldworkers commonly state that it is usually the larger troop that supplants the smaller one during encounters (Harcourt, 1992). During our four summers of research, however, Troop One was never larger than Troop Two (see Table 3.1). It is also not the case that the troop with more adult males is more dominant either. As Table 3.1 shows, there are several exceptions to that. By the summer of 1992, however, two deaths in Troop One had occurred that might have been important in the troop's fall to number two. The alpha female died in childbirth, we were told, and the beta male died. In addition, Troop Two had an additional adult male and three more adult females than Troop One by the summer of that year.

Troop Dominance and Resources

An examination of the three troops in 1990, 1991, and 1992 shows that the more dominant troop tends to control the roadside areas, especially the Garuda Area, and the Graveyard, where tourists and local villagers are most likely to provision food. Troop Three, as the most subordinate troop, ranges on the periphery of the Monkey Forest. Troop Three changed its range from the Southeast Gorge in 1986 and 1990 to the Northwest Gorge in 1992. In 1991, its range was in both gorges, running back and forth through the rice fields on the southern side of and just outside the Monkey Forest. In 1992, Troop Three stationed itself by the Pura Dalem and along the Temple Road, thereby obtaining greater access to tourist foods. In 1992, I followed a number of tourists as they walked through the Monkey Forest carrying food. The food was invariably gone before they reached the Temple Road. Food is therefore more abundant and clumped in this area. Knowing this, for example, the priestess (wife of the pemangku dasar and daughter of the pemangku dalem) would station herself in front of the Pura Dalem and would then sell more food to the tourists to feed the monkeys. In previous years, she pointedly referred to individuals in Troop Three as "skinny" and individuals in Troop One as "fat." This was one method she used in distinguishing the two troops, reflecting their differential access to resources.

The tabulated scan samples also show a large difference in frequency of feeding on tourist food. In 1990, Troop One, for example, fed on over twice the number of tourist foods, especially peanuts and bananas, than did Troop Three. Troop Three also fed on twice the number of insects than did Troop One, as if making up for the lost protein that Troop One obtained from the peanuts. Troop One was found either on or near the road on over 80 percent of the scan samples. In contrast, Troop Three's percentage was less than 40 percent. Comparing the scan samples between Troops One and Two shows that Troop One was found on or near the road two-thirds of the time, compared to one-third for Troop Two. In 1991, the proportion of human food sources in the diet for Troop One and Two was 69 percent and 73 percent, respectively. For Troop Three, the proportion was 39 percent. Ten percent of Troop Three's diet was insects compared to 0 percent for Troop One and 1 percent for Troop Two. In an examination of the diets of adult males and females in each of the troops, it was discovered that human sources make up most of the diet for the adult males in all three troops, but not for the adult females. Troop One and Two adult females had two-thirds of their diet from human sources whereas Troop Three females only had 36 percent. In 1992, a comparison of scan samples between troops showed that Troop Two received 81 percent of its diet from tourists, whereas Troop One's percentage was down to 59 percent. Interestingly, Troop Three's percentage was up to 79 percent. Troop One's insectivory jumped to 9 percent from the year before. Troop Two's insectivory was 1 percent, and Troop Three's insectivory had dropped to 2 percent.

Another way of showing the influence that food sources, especially human, have on the dominance relationships between troops is to locate on the map where these intertroop encounters occurred. In 1986, 75 percent of all the troop-troop encounters occurred where human food sources were located, that is, at the provisioning station, roads, the Graveyard, and at a small garbage dump near the Pura Dalem. This pattern continued in the years 1990, 1991, and 1992. For example, in 1992, 81 percent of all displacements were along the roads and in the Graveyard. Troop Two chased Troop One away from the central area of the Monkey Forest to the edges of it. About 40 percent of the encounters began in the Garuda Area. Three-quarters of all their encounters took place on the roads and in the Graveyard. All of Troop Two's displacements of Troop Three occurred on the Temple Road. Troop Three fled out of the Garuda Area back down the road northwest of the Pura Dalem.

There are a number of proposed hypotheses relating female relationships to resource distribution and abundance. Richard W. Wrangham (1980) proposed the most relevant model for primates. The dearth of observations on females has led to considerable difficulty in refining the model that relates the defense of clumped resources to the evolution of female-bonded troops. The research results presented here support this model. The female-bonded troops of Padangtegal are both directly and indirectly involved in intertroop competition—directly, by being involved in battles, and indirectly, by soliciting adult males in appeal aggression, for example, in scream-threats.

The behavior of these females is fascinating and needs to be integrated with the behavior of males for a fuller picture to emerge. Knowledge of the cycle state for all females in The Monkey Forest gives us a more compete picture generally unavailable in the wild. As we have shown (see Table 3.21), there are approximately equal numbers of females *in each troop* who are either mothers or estrous females. We have, therefore, a bimodal intratroop distribution of births. This bimodal distribution of births corresponds to the two tourist seasons: summer for Europeans and Americans and winter for Australians. The extra food available at those times helps fulfill the additional nutritional demands of pregnancy and lactation for these females. Furthermore, intratroop female competition is decreased when half the troop is in estrus and the other half consist of mothers and their infants. Estrous females contact and recruit males for sex and for support as defenders of the realm, while the mothers remain together and obtain resources. In addition, a lower-ranking female might benefit more from membership in a higher-ranking troop that has greater priority to resources than emigrating to a lower-ranking troop. Male integration into the troop would seem to require a year of relationships with estrous females, with the most critical relationship being with the alpha female. Finally, the data show that it is the higher-ranking females *in all three troops* who generally give birth during the drier summer months. Lower-ranking females tend to give birth during the wetter months, when conditions might presumably be conducive to greater infant mortality. The number of surviving infants during the summers of 1990, 1991, and 1992 total twenty-nine for high rankers, but only fourteen for low rankers (the bottom half of each troop). Infant mortality is greater for low-ranking mothers than for high-ranking mothers: seven for the former and three for the latter. In the summers there are over twice as many low rankers in estrus (30) than there are high rankers in estrus (13). The low-ranking adult females can still be seen to reject their old infants' (now considered to be young juveniles by us) attempts to nurse when we arrive each summer. As the grooming patterns show, the low rankers tend to associate with each other and to cycle together, just as the high rankers associate and cycle together.

Chapter 4

The Sacred Monkey Forest at Padangtegal

If I had to choose a word to describe the Balinese of Ubud, I would chose *vibrant*. They appear to be constantly energetic and busy. As outlined in chapter 1, much of this vibrance is expressed in their religion and in their arts. Such vibrancy seems to generate a powerful magnetic force that attracts more people and more energy to it. Laborers look for work, and tourists flock in. But why be so busy? What causes the Balinese to be so energetic? I propose that an explanation for this vibrancy lies in their philosophical concept of harmony and balance, called Tri Hita Karana—the three causes of goodness.

Tri Hita Karana is probably best described by the Balinese themselves. The following description of this philosophy comes from a 1996 tourist pamphlet written for the Monkey Forest:

> In accordance with the Balinese Hindu philosophy, peace and liberty are obtainable when in our lives we have performed the three harmonious relationships, known as the TRI HITA KARANA DOCTRINE:
>
> 1. God blessed life, created nature and its contents.
> 2. Nature offers subsistence, nourishment, need and activities to human beings.
> 3. Human beings have an obligation to establish traditional village structure, to build a temple to worship, to hold various ceremonies, to make daily offerings, to preserve the nature, to discuss and solve problems together.
>
> These describe the harmonious relationships of human beings with god, nature and amongst themselves. By achieving this doctrine, God will bless our karma and in our life we may avoid disaster in our natural environment.

The constant vibrancy and striving for harmony is also a check on forces of disequilibrium. As more people are drawn to the energy, more problems arise that need to be solved. Human population density increases, for example, are often cited as a root cause of habitat and species endangerment throughout the world (Czech & Krausman, 1997; Dobson, Bradshaw, & Baker, 1997; Dobson, Rodriguez, Roberts, & Wilcove, 1997). Various economic and political pressures create agricultural changes, urbanization, and tourism, which serve as leading forces of such endangerment. These forces, in turn, are also said to have an adverse effect on local communities throughout the world (Pleumarom, 1994). Some of the problems accompanying tourism, for example, are an increase in crime, an increase in environmental damage, and the loss of traditional values and beliefs. In addition, local communities generally receive very little financial reward from tourism. Bali, although certainly not immune to these problems, does continually seem to be one of the exceptions in being able to adjust to the various problems confronting it. Adrian Vickers (1994) says that the doom watchers of Bali, beginning in 1334, continue to move back the date of Bali's demise, but that "it remains in reasonable health." Philip Frick McKean (1977) and Valene L. Smith (1977) both state that tourism can strengthen traditional Bali, especially if the local involvement in tourism is broadly based. This chapter explores this broadly based or "bottom-up" approach that benefits the entire community. Once again, Tri Hita Karana is the force behind the Balinese adjustment to such problems.

The District or Regency of Gianyar

It is no accident that the cultural center of Bali is in a triangular area located between the volcanic mountains and the sea. This fertile triangle is a territorial division of Bali that is harmonious to its cosmological divisions. This middle sphere, the world of humans, is the most fertile rice-growing area and the most densely populated area in Bali. As the middle sphere of the Three Worlds in Balinese cosmology, it also lies between the godly sphere and the demonic/destructive sphere. It is the district of Gianyar that is located in the heartland of this fertile triangle. R. Friederich (1959 [1849–50]) called Gianyar one of the most fertile and best-cultivated districts of Bali in his 1848 travels on the island. In 1881, the Dutch vaccinator, Julius Jacobs (1994[1883]) remarked that Gianyar was the most densely populated realm in Bali where everyone was content. This area is also richer in antiquities than any other part of Bali; the so-called holy land of Bali lies between the sacred rivers of Petanu and the Pakerisan, the old Kingdom of Pejeng. This ancient, central region is the traditional center of Hinduism on Bali (Goris, 1960).

Gianyar is the cultural center of Bali even though it is one of the smallest in size of the eight districts on the island. Gianyar is directly connected to the Majapahit Kingdom of Java. It continued to flourish under the Dutch, when the raja, or king, of Gianyar recognized their authority. The royal families were allowed to retain their wealth and power and become patrons of the arts. In fact, it was only in the mid-1950s that Indonesia finally abolished the royal rule of the rajas (Eiseman & Eiseman, 1988).

Ubud, a Subdistrict of Gianyar

Ubud is recognized as the cultural center not only of Gianyar, but also of all of Bali (Mabbett, 1985). Ubud's antiquity goes back to the eighth century, when the priest Rsi Markandeya went to Bali and built a temple at Campuhan in Ubud (Santosa, 1985). This priest also founded Besakih, the mother temple of Bali. The name Ubud derives from *ubad*, which means medicine. Ubud was apparently rich in herbal medicine. Its high agricultural productivity is also associated with its elaborate temple ceremonies (Lansing, 1983) and its commitment to the arts. While most of the 2.9 million population (1996) of Bali are engaged in agriculture, it is instructive that the percentage of people in Ubud who are artists, craftsmen, and sculptors totals 24 percent of the population in comparison to farmers, who make up 20 percent of the population (1990 census). Vickers (1990) says that most of the people of Gianyar rely on tourism for their income. The attractions that draw these tourists to Ubud have given rise to its local designation as a "honey area." Over a million tourists are estimated to visit Bali every year (Picard, 1990b).

Monkey Forests

It is in the forests, ravines, and cliffs of Bali where monkeys can be viewed. Such areas are natural sanctuaries where monkeys can flee to safety from their potential predators, such as humans and perhaps dogs. Balinese hunt monkeys using dogs and rifles, as I have seen in the riverine gorges near Bangli (Wheatley, Fuentes, & Harya Putra, 1993). Other sanctuaries are in sacred village areas, often surrounding temples. These "cultural" sanctuaries are not only an important part of Balinese heritage but also an important part of everyday life. Temple festivals are regularly held for the villagers and the gods in such areas. A Balinese temple is more than just a collection of pagodas and pavilions. The area enclosed by temple walls and the forest area surrounding it is sacred. These temples and the forest are essential for renewing contact with the spiritual world. The activities associated with these areas are essential in maintaining harmony between humans, nature, and the cosmos. Not only are ancestral spirits and gods given offerings and prayers, but also the spirits of trees and statues in the Monkey Forest are given offerings and prayers by the pemangku and local villagers. The various sacred masks, for example, of Rangda (see Plate 1.4) can only come from the wood of sacred trees in this forest. In addition, local villagers provide offerings, conduct graveyard ceremonies, and go for their evening strolls in the forest sanctuary on a daily basis. In fact, my friends used to find it hilarious to ask if I'd just been out for my evening walk, after I had just spent all day studying monkeys in the forest.

The presence of sacred forests is a demonstration of the harmonious coexistence of humans and nature. The idea of sacred forests probably derives from India, where they have existed all over the country for thousands of years (Singh, 1997) and are often associated with water (Gadgill, 1985). Such reserves receive special protection from the local community, and they are often the abode of or associated with particular deities. The deities look after the welfare of the local community and help solve its problems. The Balinese traditionally keep these forests free of all exploitation for fear of offending the deities. Many of the plant and animal species are considered sacred, especially fig trees such as *Ficus religiosa* and monkeys. According to M. Gadgill (1985), monkeys are never hunted, even when they raid crops, perhaps because of their close resemblance to man. An example of the extent of the protection given to certain species comes from an account of 363 members of the Bishoi Hindu sect who sacrificed their lives in A.D. 1630 to prevent the King of Jodhpur from cutting down sacred trees. This example is vivid testimony to the relevance of building conservation from the ground up by incorporating traditional indigenous beliefs and practices.

The Monkey Forest actually belongs to the Desa Adat of Padangtegal, the religious/customary unit of about 500 villagers that manages the Monkey Forest. Hildred and Clifford Geertz (1975) define the desa adat as a group of cooperating congregations that support the three temples: Pura Puseh, or origin temple of the village; Pura Bale Agung, or great council of the gods; and the Pura Dalem, or death temple. The overall function of this religious community is to maintain the cosmic balance of the village and to assure the health and welfare of its inhabitants "by worshipping the divine powers and exorcising the demonic ones" (Swellengrebel, 1984). Padangtegal is a village whose name means "dry" field or nonirrigated agricultural land, although most of the rice crops are now intensively irrigated.

The location of the Monkey Forest follows the typical Balinese pattern of the threefold division of space, called Tri Angga. The Pura Dalem and the Monkey Forest are potentially polluting and dangerous, because of the graveyard among other things. It is therefore located toward the sea or away from the sacred volcano, Mount Agung and south of Padangtegal and Ubud. The other two temples, the Pura Puseh and the Pura Bale Agung, are respectively located centrally and northerly to Padangtegal. The Pura Dalem is associated with Siwa, the destroyer, and his wife in her demonic form of Durga. There are actually three temples as well as a graveyard located in the Monkey Forest. The largest temple is the Pura Dalem and again, it is the most southerly of the other two temples. The cremation temple is toward the volcano, and a bathing temple is toward the west. The Tri Angga applies not only to the location of the temples of a village along the mountain-sea axis but also to many other things, such as to their housing, and to the human anatomy of leg, body, and head. In fact, this concept is a derivative of the philosophy of Tri Hita Karana.

Tri Hita Karana

The philosophical concept of Tri Hita Karana has received little attention by anthropologists. It is a concept of balance and unity that is sometimes cited as important in the health of individuals (Jensen & Suryani, 1992; Muninjaya, 1982; Wikan, 1989). It is, however, more than that. It is a philosophical method of viewing the Universe with the purpose of achieving a balance. Eko Budihardjo (1986) mentions it as one of the motivating factors for the arts of Bali. It is specifically mentioned as a principle of management at the Monkey Forest. Its derivatives are even expressed in architecture.

Tri Hita Karana means the three causes of goodness. *Tri* means three. *Hita* is the Sanskrit term for welfare, benefit, or goodness, and *Karana* is the Sanskrit term for the supreme cause or act of making, producing, or affecting

(Zoetmulder, 1982). This philosophy states that the three elements of the world that come from God and infuse the Bhuwana Alit (the microcosmos or person) and the Bhuwana Agung (the macrocosmos or the natural world) are: *atma* (soul); *prana* (breath); and *sarira* (body). In the Bhuwana Alit, these three elements make up the human world as follows: The atma is the essence of God, causing human life; when it infuses the sarira, or physical body, an organism is created. From the unity of body and soul comes energy or power (prana) that shapes the three supernatural human forces of *bayu-sabda-idep*, or force-word-thought. In the Bhuwana Agung, these three elements are *paramatma* (the soul of the universe), the *panca Mahabhuta* (the five coarse elements), and prana, which is the sum of all types of power and energy in the universe (Kaler, 1983). The panca Mahabhuta can be identified with water, wind (bayu), air or sky, light or heat, and the earth or land (Budihardjo, 1986; Hooykaas, 1974). It is the balance or unification of all these elements that brings about happiness. The opposite, of course, is also true, and the disturbance of any of these elements can result in disorder or disease.

Tri Hita Karana thus underlies both the microcosmos and the macrocosmos and unifies such opposites as male-female and life-death. Such a unification of apparent opposites into one is referred to as rwa bhineda, and this unification will lead to moksa, or heavenly bliss (Budihardjo, 1986; Zurbuchen, 1987). The synthesis of opposing forces is said to occur in the middle, at which all is harmoniously united. This position is the emptiness and calmness of zero, where all aspects of life are perfectly fused (Laksono, 1986).

Bali is well known for its emphasis on balance, order, and equilibrium, both within the cosmos and within the individual. As an agrarian society, the Balinese strive to continue the cycles of nature—the rainy/dry season and the cycles of life—birth, growth, and decay. Their ceremonies help maintain harmony between an individual and his or her environment (Muninjaya, 1982). Perhaps one of the best examples of this balance is one of the life-cycle rites, namely tooth filing. Generally all Balinese should have their six top front teeth filed by a Brahmin priest after puberty and usually before marriage. These teeth, four incisors and two canines, are symbolic of animality, and they are filed flat, especially the canines, so that they do not protrude beyond the tooth row. It is interesting that only the top teeth are filed, while the lower teeth are untouched. Anthony Forge (1980) says that such cultural modification is necessary to produce a human, but that the lower teeth are untouched in order to preserve some of the animal passion typical of nature. Traditionally, the teeth were also blackened, further distinguishing them from the white fangs of animals. Forge also points out that the only animal depicted with the correct or "natural" occlusion of canines is the pig. The carved masks of pigs have the lower canines in front of the top canines, which is the correct configuration, in contrast to the Balinese depictions of monkeys or other animals in whom the reverse is the case. The filing of the teeth is said to help rid an individual's spirit of the six weaknesses: lust, greed, anger, laziness or intoxification, jealousy, and confusion, infatuation, or folly. Tooth filing is important to an individual not only in this world but also in the next (Eiseman, 1989) (see Plate 4.1).

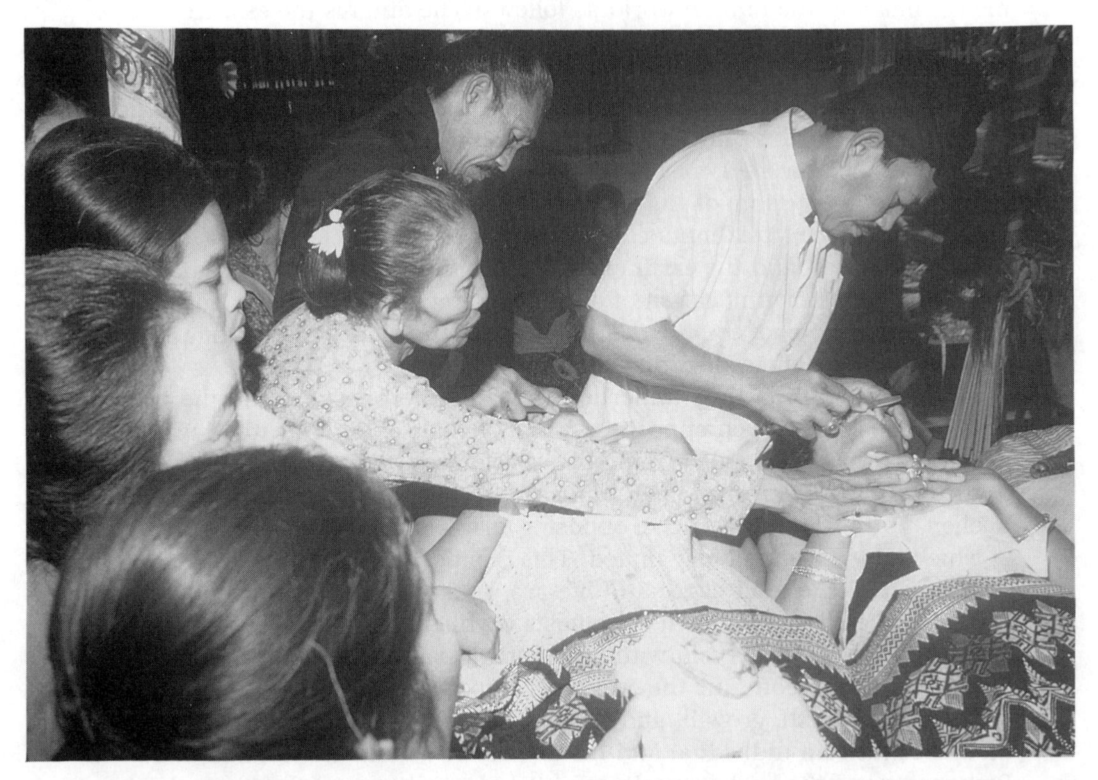

Plate 4.1. *Brahmin priests file down the top front teeth, especially the canines, of these women in Padangtegal. Men also have their teeth filed down.*

The unity of microcosm with macrocosm is, says Christiaan Hooykaas (1970; 1974), a basic and beloved piece of Balinese philosophy. He cites one of the most striking examples of such unification in the ancient manuscript of the Purvaka Bhumi, The Beginning of the Earth. This unification is exhibited in the *kanda mpat* or four spiritual siblings, which are incarnation of the god Siwa. The four siblings are born along with every baby. They are physically manifested as the placenta, the amniotic fluid, the blood, and the vernix caseosa, which is the whitish/yellowish coating on a newborn. Spiritually, they are identified with various gods (Brahma, Wisnu, Isvara, Mahadeva), colors, directions, mantra syllables, texts, temples, dances, letters of the alphabet, weapons, and mounts. They reside in every person such as in the liver (Brahma), heart (Isvara), kidneys (Mahadeva), and bile (Wisnu). The pandasars (clown-servants) also belong to these locations. Twalen, for example, belongs to the liver. They can help you throughout your life provided they are treated properly. If neglected or ignored, however, they can turn into *bhutas*, or mischievious spirits, and become followers of Rangda, the witch, and cause illness. Although the four siblings are exclusive to Bali and alien to Hinduism, Brahmin priests invoke them during life-cycle rites.

In general, the maintenance of the Balinese ideal of balance, order, and equilibrium within the cosmos and within the individual is through religion. The causes of imbalance are, likewise, also sought in their religion. For example, a common form of discord is between a man's mother and his wife, who all live in one compound. In the Western view, there may be many reasons for this discord, all of which may, in fact, be valid. Such things as personality conflicts, jealousy, or favoritism can be reflected in differential workloads or differential materialistic and social rewards. A common Balinese response to this problem, however, is to seek the explanation of the discord through divine guidance. A spirit medium will be consulted, and the ancestors/gods will inform the parties of their displeasure. Perhaps a particular ceremony has been omitted, and certain corrective rituals and offerings will need to be made. The solution is to correct the imbalance between various individual differences by bringing the cosmos in line with the individual world and thus achieving harmony. Another common example of the maintenance of order is to ritually placate various *buta-kala*, or noxious spirits, in order to effect a cure of an illness. Such spirits have temporally entered a person, and those spirits need to be converted to their divine form, which will reorder the cosmos (Howe, 1984; 1989). Once again a balance is struck between the cosmos and an individual person. The interrelationship among such entities as the cosmos and people and their harmonious ideal is the philosophy of Tri Hita Karana.

The Questionnaire and Community Attitudes

Before I discuss how the concept of Tri Hita Karana was used at the Monkey Forest, I need to set the community stage. When I first arrived in Ubud, in June of 1986, there were only a few hotels and shops on the northern edge of the forest and none on the southern edge within a kilometer of the forest. The roads were unpaved, and local villagers grazed cows in a grassy field just north of the forest. Relatively few tourists visited the Monkey Forest, only 800 a month according to the registration book that tourists signed when they entered the forest.

By 1991, both roads to the forest had been paved, and there were hundreds of tourist shops and dozens of hotels. The grassy field where cows grazed and monkeys rested was now a parking lot. By 1992 a dramatic boom in tourism was noticeable. About 16,000 tourists visited the forest in an average month.

Over the years, I have been interested in how local villagers viewed the Monkey Forest and tourists. I talked with many villagers, including farmers, priests, artists, retired people, shop and hotel owners, and local guards at the forest (see Plate 4.2). In 1992, I Made Sada Artha, a young Balinese man,

Plate 4.2. *A local Monkey Forest guard holds a dead civet cat.*

assisted me in obtaining further information on the attitudes of Balinese living or working in the area. He spoke Balinese, and he had previous experience interviewing villagers for his thesis at a university.

My study group designed a questionnaire soliciting information about Ubud and the Monkey Forest (see Table 4.1). We received responses from 136 people, and I analyzed the variables statistically using Stat Pac, Inc., a data manager and analysis package for the computer. I Made Sada Artha also obtained information from the 1989–1990 census from the office of the Lurah, or village headman, in Ubud. The census figures for the village of Ubud show a population of 8,150 people as of March 31, 1990. This is an increase of 313 people from the census of 1988, which indicated a total of 7,837 people. Between 1988 and 1990, additional sections were added to Padangtegal and Ubud. According to the census, there were only twenty-four Muslims and eleven Buddhists, with the rest Hindus (99.5 percent). The occupations were: farmers, 20 percent; merchants, 12 percent; government employees, 12 percent; laborers, 11 percent; artists, 10 percent; private employees, 9 percent; craftsmen, 7 percent; and sculptors, 7 percent. The results of the questionnaire to the villagers are presented in Table 4.1. The 136 questionnaires represent almost 2 percent of the entire population of Ubud. An examination of the data shows a number of significant differences using the chi square test.

The Monkey Forest is clearly important to the Balinese culture. Ninety percent of the villagers indicated this. Even non-native Balinese who are predominately Muslims said so (85 percent). The Monkey Forest is not just for tourists. Nearly half of the respondents indicated that they visited the forest three or more times a week. Twenty percent visited it daily. The only category of people who infrequently visited the forest were shorter-term/temporary residents. Affirmation of the sanctity of the forest can be seen in the near consensus that shops and hotels were too close to the forest. Locally born Balinese were nearly unanimous in answering yes on this question (98 percent), while the percentage of non-Balinese answering yes was 63 percent. Most of the

Table 4.1. **Villagers' Questionnaire Results**

1) **Total Number**	**Under Age 25**	**Over Age 25**	
136	44%	56%	
2) **Total Number**	**Male**	**Female**	
135	74%	26%	
3) **Total Number**	**Born in Ubud**	**Born in Bali, Not Ubud**	**Foreign-Born (Not in Bali)**
133	60%	26%	14%
4) **Total Number**	**Length Residence in Ubud Less than 5 years**	**Length Residence in Ubud More than 5 years**	
132	38%	62%	

5) **Occupation (Total Number = 129)**	**Percent**
a) hotel and restaurant workers	20%
b) artists	18%
c) farmers	14%
d) silverworkers	10%
e) merchants (hawkers)	10%
f) self-employed	8%
g) woodcarvers	8%
h) shopkeepers and government workers	4%
i) students	3%
j) garment workers	4%
k) doctors	1%

6) **Religion (Total Number = 136)**	**Percent**
Hindu	87%
Islam	12%
Christian	1%

7) **Times per Week Visit The Monkey Forest (Total Number = 116)**	**Percent**
a) never or very infrequently	12%
b) once a week	15%
c) twice a week	25%
d) 3–5 times a week	28%
e) 6 or more times a week	21%

Question	**Total Number Questioned**	**Percent**	
		Yes	**No**
8) Income from tourism	135	66%	34%
9) Family residence in Ubud	136	53%	47%
10) Work in Ubud all year	113	73%	27%
11) Monkey Forest important to Balinese culture	125	90%	10%
12) Monkey Forest important to your success	125	68%	32%
13) Advertise the Monkey Forest for tourism	129	78%	23%
14) Food prices higher in Ubud	115	63%	37%
15) Land and house prices higher in Ubud	122	99%	1%
16) Shops and hotels to close to the Forest	133	90%	10%
17) Shops and hotels contribute to conserve the Forest	103	78%	22%
18) Too much Western influence in Ubud	130	40%	60%
19) Too many tourists	129	98%	2%
20) Easy to marry foreigner	115	93%	7%

Total possible completed questionnaires = 136.

people questioned felt that shops and hotels should contribute to the maintenance of the forest. The people we questioned said that there were too many monkeys (99 percent), but that they had not become more aggressive (75 percent).

Another reason why the Monkey Forest is important to the Balinese is clearly economic. Two-thirds of the questionnaire respondents said that the Monkey Forest was important to their success. This is the same percentage reporting that their livelihood depended on tourists. Interestingly, only non-native Balinese reported that their income did not depend on tourists, and this percentage was by a two-thirds margin. A significant number of respondents in all occupations (except farmers and silverworkers) said that the forest has helped their success. Three-quarters of the respondents said that the forest should be advertised to tourists even though they felt that there were too many tourists already.

The most remarkable part of these results is documentation of a recent influx of non-Balinese. Eighteen of the respondents said they were born outside Bali, of whom seventeen are Islamic. This number is almost as high as the total number of Muslims in the whole village according to the census. Most of these Muslims are recent arrivals and silverworkers. Only three Muslim respondents answered the question "Do you work here all year?" This question is obviously a sensitive one. Nonresidents who work year round are required to check in with the local authorities. Muslims and individuals of other religious groups are not a member of any Hindu desa adat. A recently constructed gate with a lock is located in front of a bridge leading down to a temple and bathing facilities in the gorge. Above this gate is a sign in Indonesian stating that this is a holy Hindu area and that it is forbidden to pollute it with garbage and human waste. Further inquiries revealed that it was alleged that some Muslims working on construction jobs in the area, especially on a new Javanese-financed hotel next to the Monkey Forest, had been polluting the river. As a check on the ease of integration of non-native Balinese at Ubud, we asked if it was easy to marry a foreigner. Almost everyone said no.

Another element of this recent influx is the nonlocal Balinese who have obtained jobs in Ubud. Almost 30 percent said that they did not work year round in Ubud. Most of them are helpers in hotels or restaurants and most are males (68 percent). These people are mostly Hindus, however, unlike most of the non-Balinese immigrants. Nevertheless, a common local complaint was that the newcomers were profiting from the Monkey Forest but that they were not contributing to Padangtegal. They were apparently becoming members of Ubud rather than Padangtegal. We were also told that traditionally only females came into the village at marriage, but that now, mostly males came into the village at marriage. Local people were concerned about jobs going to nonlocal people. For example, in 1986 nonlocal people were selling handicrafts in the Monkey Forest, but by 1990 they and almost everyone else had been forbidden to do so. Some jobs were not attractive to the local people, however, such as building roads. Workers from other villages, mostly women, were brought in to do the work.

The last major area of information that we were interested in was how the villagers felt about tourism. Nearly everyone we questioned said that there were too many tourists (98 percent). Interestingly, 60 percent of the people questioned said there was not too much Western influence in Ubud. Breaking down this result by religion, age, and sex, however, shows that not only do a majority of Muslims feel that there is too much Western influence but also most females and people under the age of twenty-five years. People were nearly unanimous in stating that the prices of land and houses were higher in Ubud than in other villages. Of special concern was the influx of foreigners (both Westerners and Javanese) who were financially supporting the building boom of new hotels and restaurants. People also stated that Ubud had higher food prices than other villages. A common complaint was that foods normally available during the peak of tourist season, generally in August, were very difficult to obtain. One interesting result is that 61 percent of the older long-term (over fifteen years) residents of Ubud said that food prices were not higher in Ubud than in other villages.

The effects of tourism in Ubud are not unique. Even more dramatic change occurred at Kuta, a small village along the beach (Hussey, 1989). Between 1970 and 1984, for example, Kuta changed from a sleepy fishing village to a significant tourist destination. From a recorded 6,000 tourists staying in Kuta in 1972, the number rose to over 60,000 by 1984. The number of *losmen* (small motel-like rooms), hotels, and restaurants likewise increased to accommodate the tourists. The price of rice fields rose an insignificant amount compared to the price of beachfront and core roadfront land in the village (by a tenfold factor). At first, the local people predominated in the prosperity, but later, Indonesians from Java, and Westerners began to invest until, for example, by 1983, almost half of all bars and restaurants in Kuta were owned by foreigners. The environmental effects of tourism were also notable at Kuta. Air and noise pollution from traffic congestion, destruction of coral beds, and subsequent beachfront erosion, and trash accumulation, especially plastic bags and straws, were among the worst environmental problems. Another problem was the introduction of illegal activities such as crime, drugs, and prostitution. One interesting point mentioned by Hussey (1989) is that the tension between the provincial police and the villagers had increased with periodic accusations of corruption because the village no longer provided for security as was traditional. Despite these serious problems, however, the local villagers say that they are adjusting and they deny that traditional Balinese life has been greatly affected (Setia, 1994).

Ecotourism has generally impacted communities adversely throughout the world (Pleumarom, 1994). Margaret Kinnaird and Timothy O'Brien (1996) found that the local community near the Tangkoko Dua-Sudara Nature Reserve was not significantly profiting from the ecotourists visiting the reserve, especially to see the local macaque, *M. nigra*. They found that only a fraction of the income (2 percent) generated from ecotourism went to support management programs for the Reserve and that the population of macaques had declined 75 percent over the last fifteen years. In Ubud, however, the villagers formulated a plan to attempt to contain these problems, to curb outside development and to use ecotourism for the benefit of the community. For example, organic farming has been started, a newspaper has been published, and an information center has been established by the local people (Pleumarom, 1994).

The Sacred Monkey Forest: The Managerial Committee and Conservation

A similar process of conservation awareness and activism is occurring in Thailand. At the core of this movement are the so-called Buddhist ecology monks, who see the cause of their countries' deforestation and pollution as a moral issue (Darlington, 1998). By publicly engaging in tree ordinations, the ecology monks are combating greed and ignorance by reminding people to treat nature on equal terms with humanity. The monks perform skits on how to preserve the forest and they sanctify seedlings prior to planting. They then wrap monks' robes around all the large trees that are threatened. After the ceremony, they put up signs on the ordained trees that read "To destroy the forest is to destroy life" (Darlington, 1998). The cutting of any trees or killing any wildlife is forbidden. The village headmen also drink holy water and pledge to protect the forest. A violation of this pledge might affect their life for all eternity, including their rebirth.

In Bali, by 1990, the change from a small village atmosphere of 1986 to a major tourist attraction at Ubud and Padangtegal was obvious. One section of Padangtegal, had grown in population so much that it had to be divided. Hanuman Street from the city of Ubud to the Monkey Forest was now paved. Local villagers now talked about the huge influx of tourists, attracting all kinds of people. Padangtegal was "more green" than other villages. The rice fields were more prosperous. The people of Ubud also mentioned more problems now that accompany tourism. They worried that the more Western lifestyle of tourists might influence their own children. They mentioned theft and pollution for the first time as potential problems.

In 1990, the Wenara Wana Padangtegal Managerial Committee was established by members of the desa adat in the outer courtyard of the Pura Desa (village temple) in Padangtegal. The purpose of this committee is to manage and conserve the "tourist object" Wenara Wana. The Sanskrit *wenara* means "forest animal or monkey," and *wana* means forest (Zoetmulder, 1982). The previous name for the Monkey Forest was, in Balinese, Alas Pemaosan. The name change from Balinese to Sanskrit is a significant one. Sanskrit is perceived as more than just evoking a more traditional image of the past. It is a sacred language and a vehicle of potent information to stand the test of time and modernization. The name change reflects an elevation in status for the Monkey Forest, and it reinforces the sanctity of the forest. Asked why the name was changed and whether the monkeys were sacred, an elderly man told me the following story. Upset that monkeys were always harvesting the forest's coconuts and other foods, the local people made a determined effort to get rid of the monkeys in the 1930s. Some of the animals fled to Kedisan, and for eight months there were no monkeys in the Monkey Forest. Gradually, however, people began to see more and more monkeys in the forest. Clearly, said the old man, God wanted them here. Interviews with various villagers showed that they felt that saving monkeys helped to save their culture and that the forest served a valuable religious function.

I obtained copies of the yearly programs of this committee, and my Balinese colleague, Harya Putra at the University of Udayana in Denpasar, translated them. There were two reports, 1991–92 and 1992–93. The first year's program discusses some of the problems of the Monkey Forest. The forest and land was deteriorating because of erosion, pollution, and other problems. Although tourists came to see the monkeys, their density had become too high, and other animals had become rare. The previous workers' duties and wages were inadequately specified, and the collection of tourists' fees was not regulated. Documentation of income and expenses was inadequate. The workers' wages were believed to be too high—up to 40 percent of the total—and none of the income went to conservation. Tourist facilities such as rest rooms, parking areas, rest areas, and footpaths were also inadequate. The second part of the 1991–92 program states general categories of conservation, fees, visitor safety, maps, and informational meetings for members of the Desa Adat of Padangtegal. Specifically, they proposed to change the feeding areas for the monkeys to avoid further damage to the area and to provide water for them. Erosion was to be controlled by building terraces and growing plants along the terraces, and by planting 5,000 seedlings of plants that could be used for ceremonial purposes. They also hoped to save species such as the jungle fowl and iguana by forbidding all hunting. Workers needed to be hired to clean the area and to provide trash cans. They proposed to add a hectare of land to increase the size of the forest. Better security was needed; more guard posts, parking lots, and traffic signs were necessary, and unauthorized vehicles were not allowed to enter the forest. They proposed to provide electrical power to the Pura Dalem and to have lighting at other locations. Financial support was also to be sought from businesses and the government. Especially important was the proper recording of the daily number of visitors and the income generated. All of the benefits from this new management were to help preserve the tradition and religion of the Desa Adat Padangtegal. The 1992–93 program of the Wenara Wana Padangtegal Managerial Committee states that the environment and nature, or flora and fauna, as well as the artistic and other cultural activities of the community are integral components of their Hindu religion. All of these components should be utilized as important resources in promoting tourism for the equitable benefit of the people, especially the Desa Adat Padangtegal.

Most importantly, the program states that the management of the forest is based on the philosophy of Tṛi Hita Karana. That is, the management must strive to balance or to develop equilibrium between the creator, the people in and outside the community, and the environment. The committee, therefore, proposed to strengthen the religion, conserve the forest, and to increase the economic benefits for the Desa Adat Padangtegal. The economic benefits of tourism can significantly raise the per capita income, and this can be as equitable as possible, that is, not corrupt, only if every visitor to the Monkey Forest has paid for a numbered admission ticket. The daily receipts are collected and tabulated to ensure proper accounting. This "open management system" is in marked contrast to the previous "closed management system." Previously, the recording of income and expenditures was inadequate, with expenditures for wages as high as 40 percent of the gross income. The remainder of the income went for other purposes with nothing for conservation. In 1992, each tourist paid 500 rupiah entry fee (about 25¢ U.S.). According to the treasurer of the managerial committee, the total receipts from tourists over the year amounted to 57 million rupiah, or about $30,000 U.S.

The income from tourism is roughly split into thirds, with one-third going to conservation and the rest toward the maintenance and ceremonies at the three temples. Previously, for example, all members of the desa adat had to pay or donate money for offerings and ceremonies, and it was sometimes difficult to collect. Now, the money came from tourists, and as the head of the managerial committee told me, "The money goes to God." Individuals who did not belong to the desa adat had to pay for these services. The wealthiest of the three temples is the Pura Dalem in the Monkey Forest (see Plate 4.3). In the past, this temple was just for commoners, but now Brahman priests also go there. This temple was being renovated with new electrical power lines, marble and gold leaf. The temple also owns rice fields and makes income from the sale of the rice. The community would also benefit from new job opportunities such as construction projects in the Monkey Forest and as tour guides and translators. Religious training of the younger generation was also to be intensified to minimize the negative effect of tourism. The villagers also wanted to restart their bank.

The conservation goals of the 1991–92 program were in place by 1992. Stone terraces and footpaths had been built for controlling erosion, and trees had been planted in a grassy area to increase the size of the forest. The few surviving trees not destroyed by the monkeys, were *gamal* trees, *Gliricidia sepium*, whose leaves can be used as cattle feed. A small pond provided water to the monkeys. Trash bins were also provided. Those donated by a local restaurant had its name on them. A toilet costing 2 million rupiah was being installed, guard posts and ticket booths were conspicuous, and a parking lot and electrical power had been installed. Unfortunately, several monkeys were electrocuted when they got into the power lines. Unauthorized vehicles were not allowed into the forest. Food for the monkeys was now only sold at the entrances except for the priestess, who was allowed to sell food in the forest.

The man most responsible for the reorganization of the management of the Monkey Forest explained some of the dangers of tourism and corruption by telling me a story about the monkeys in the Monkey Forest. When he was a boy, he guarded his family's fields against monkeys who would raid them. He was impressed with how smart monkeys were. They always figured out when he had left to eat lunch, for example, or when he left his shirt as a scarecrow.

Plate 4.3. *The author and his wife pose with a village priest in the outer courtyard of the Pura Dalem. Photo by Earthwatch volunteer Lou Harrell.*

He watched how the male leader would maintain his vigilance for farmers, who would chase the troop with sickles. It was only after the rest of the troop had fed that the alpha male would come down from his lookout tree to feed. Today, the man watches the alpha male accept food from tourists. Rather than wait until after the rest of the troop feeds, however, he is the first to feed, and he chases his wives and children away from the food until he has had his fill. We have much to learn from these monkeys, said the man. The monkeys had become corrupt, like the farmers of today who sell their land for quick profits, leaving hotels and restaurants financed by nonlocal people. Unfortunately, said the man, the money is soon gone, perhaps gambled away, and the family is left destitute and without its livelihood.

Anthropologists are in a position to work effectively with local people to solve certain problems. The importance of local people in conservation efforts is now becoming more apparent (McNeely, 1992; Orlove & Brush, 1996). In their studies on rhesus macaques in India, Charles Southwick and M. Farooq Siddiqi (1985) stress that the major factor responsible for the increased populations in a few areas was the protection local people provided these monkeys because of their religious beliefs. The increasing rate of habitat destruction is causing worldwide declines in nonhuman primate populations. While the conservation efforts of one village at a Balinese monkey forest may be insignificant to the overall fate of many endangered fauna and flora, it is an indication that economic and/or religious activities can reverse declining populations of primates and play a major role in conservation (Grove, 1992). As the previous section pointed out, the financial resources of tourism at the Monkey Forest are supporting the conservation and religious goals of the Monkey Forest Managerial Committee.

Unfortunately, it is not enough to charge entry fees to tourists coming into the Monkey Forest and to use that money for conservation projects. The great popularity and uniqueness of such forests, where tourists can view free-ranging monkeys from only a meter away, also has the potential for danger. The most obvious danger is that the familiarity and ease of contact between monkeys and people will result in problems such as biting and/or the transmission of diseases. Some of the many guidebooks to Bali describe monkeys in very negative terms. The monkeys of Sangeh, according to Hedi Hinzler (1990), are "a nuisance, for they attack visitors and steal their spectacles, jewelry, watches and handbags and make life impossible for souvenir vendors in little shops close by." Michel Picard (1990b) has also called the Ubud monkeys "daring rascals" and "rapacious thieves and dangerous if provoked." It is obvious that if the monkey forests are to continue to exist, potential problems between monkeys and tourists will need to be avoided. If the monkey forests cannot be wisely managed, then monkeys will be eradicated. For example, the Singapore Botanical Gardens had a large population of monkeys, but problems between them and visitors forced the authorities to totally eradicate the monkeys in the early 1970s. The anthropological primatologist can play a very useful and significant role toward assisting management in controlling monkey-tourist interactions.

Over the course of my research during the summers of 1986, 1990, 1991, and 1992, the extent of monkey-tourist interactions became more noticeable, and I have worked more closely with the local people. In 1986, for example, I documented the important effect of human food sources on monkey-monkey aggression. Monkey aggression on tourists was infrequent. Over the years, however, the incidence of monkey bites on tourists increased. Mabbett (1985) reported one example when a monkey bit a tourist. This incident was broadcast on Australian television in 1984 and created quite a furor. The program had given an inaccurate account of the many dangers of Bali, suggesting that many tourists are bitten. Worried about the negative effect this television program would have on tourism, both the Australian and Indonesian airlines presented a more accurate and detailed account of some of the potential problems of tourists.

The amount of food provisioning was increasing over the years. In 1986, only the top-ranked troop appeared to obtain most of its diet from human provisioning. By 1992, however, all three troops were doing so (see chapter 2). This provisioning was a source of competition among all three troops, and the rate of violent encounters between troops was increasing (see chapter 3).

Intratroop aggression was also higher when tourists fed monkeys than when they did not feed the monkeys. Monkeys threatened each other significantly more (noncontact aggression) and hit or bit each other significantly more (contact aggression) in the presence of tourists with food (t-statistic, $p<.001$). Both contact and noncontact aggression occurred significantly more on the road than off the road, on the road being where most human food sources are located (t-tests, $p<.001$). Monkeys were thus more aggressive around locations where food and tourists were. The animals relaxed and groomed each other in areas off the road and out of view of tourists. They groomed each other, for example, significantly more when tourists were absent (t-test, $p<.005$). The presence or absence of tourists did not affect the frequency of play, however, and the young monkeys tended to play more on the road. Such results are fairly typical. Studies on rhesus monkeys, for example, have shown that feeding increases frequencies of aggression two to six times above those of nonfeeding periods (Southwick, Siddiqi, Farooqui, & Pal, 1976).

Another important factor in monkey-tourist interactions is the monkey population size and density. Over a five-year period, the population doubled in size. In 1968, the population size was about twenty-three, according to local villagers. Provisioning was apparently started in 1976, not only to enable tourists to see the monkeys but also to keep the monkeys out of the farmers' fields. The total number of monkeys in the forest at that time was about twenty-five, according to the local villagers. In 1978, a Japanese researcher counted thirty-one animals (Koyama, Asnan, & Natsir, 1981). Table 3.1 shows a dramatic rise in population to sixty-nine individuals in 1986 and an almost doubling of the population between 1986 and 1990. An examination of the age structure in Table 3.1 shows a decreasing percentage of immatures in the population down to 36 percent in 1992. Immatures refer to both infants and juveniles. Together with the decreasing natality (infants/adult female) percentage of 42 percent, these figures predict a declining population in the future. Working on rhesus macaques, Southwick and Siddiqi (1977) indicate that any proportion of immatures below 50 percent suggests a declining population.

The increase in population size has greatly increased the density of animals which, by 1992, was about 1,300 per square kilometer. A map of the forest was made (see Figure 2.2). The forest is about four hectares in size. The increase in density probably helps account for the increase in troop-troop aggression, which in 1991 was at a rate of 1/6.15 hours of observation. This rate is more than double that of 1986, which was one incident of aggression per 15.5 hours of observation. Both the increases in population and density may account for the increasing infant mortality rate. The proportion of infants in the population during the last few years dropped to about 12 to 13 percent.

The rapid increase in the monkey population was a serious concern to the local people. Our questionnaires revealed that 99 percent of the local people whom we interviewed thought there were too many monkeys. Members of the Managerial Committee also asked us what we might recommend to relieve the overpopulation. Among the ideas we discussed were trapping and translocating the most subordinate troop (Troop Three), which raided the farmers' fields most, or the juvenile/subadult males that we called the Garuda Boys, who were very aggressive to tourists, or even implanting birth control devices. The committee flatly rejected these ideas. A number of years earlier some monkeys were trapped, and blood samples were taken. After they were released, the monkeys became very aggressive toward people for months. Trapping destroys the balance among God, man, and nature, according to the committee. The members of the committee were especially interested in the high infant mortality and kidnapping in the forest because these events were "more natural." The monkey population has increased to 155 animals according to a 1996 brochure now handed out to tourists.

Another factor accounting for more monkey-tourist interactions is the rapidly increasing number of tourists. The large increase in tourists was enough to keep the greatly increased monkey population at about the same level of percentage of their diet from human sources. Estimating the diet of Troop One from an analysis of the 1992 scan samples shows that about 56 percent of their diet came from human sources. That is, 27 percent of their diet came from bananas, 25 percent from peanuts, and 4 percent from offerings and other food provided by tourists. Interestingly, the feeding of bananas by tourists appears to have replaced the provisioning of sweet potatoes by local villagers.

By 1990 and 1991, it was apparent to both us, the anthropologists, and to the local villagers that the interactions between monkeys and tourists were changing in both quantity and quality. We found ourselves working together more and more. In 1991, I discussed my research more extensively with members of the managerial committee. I gave them copies of my research conclusions and drafted guidelines in English for tourist safety in the Monkey Forest. I also gave them copies of the maps I used to plot the locations of monkeys and other things in the forest. These guidelines and maps were displayed in the Monkey Forest the next year. The signs warn tourists not to let monkeys touch them and to feed monkeys from a safe distance, not making contact. Tourists were warned not to grab a monkey and that if one climbed on him/her, the best thing to do was to drop the food and walk away until the monkey jumped off. The maps were prominently displayed at two entrances to the forest showing the locations of roads, temples, and so on. It is instructive to note that in 1986, the local villagers were not very much interested when I suggested putting up a few signs for tourists.

In 1992, I decided to focus on the effect of tourists on monkey aggression. We interviewed tourists, filled in questionnaires with their responses, followed them in the Monkey Forest, and recorded their interactions while they fed the monkeys. We also decided to include tourists at another monkey forest about 17 kilometers from Ubud, called Sangeh. The latter is more famous in some ways than Ubud because of its more ancient and sacred character. Mabbett (1985) says that the trees in this forest are actually dipterocarps rather than the often-stated nutmeg trees. We thought it would be interesting to compare the two monkey forests with regard to the monkey behavior, the management style, and the interactions of tourists and monkeys. We therefore devised questionnaires and interviewed tourists in English, German, French or Indonesian as they left the forests of Padangtegal and Sangeh. Earthwatch volunteers assisted my two student assistants, Katy Gonder and Tripp Holman, and me in this task. The questions and results are presented in Tables 4.2 and 4.3. Both tables summarize the data for both sites combined and calculate t-tests on the "yes" and "no" percentages (in the column labelled "Both Sites"). In the last column of both tables, the data are broken down by site with chi-square tests comparing the two sites. We filled in ninety questionnaires at Padangtegal and thirty-three at Sangeh. We have less data for Sangeh than Padangtegal, primarily because the monkeys at Sangeh were fairly dangerous and attacked us. Some of these results are presented in more detail in Wheatley and Harya Putra (1994a; 1994b).

Table 4.2. **The Results of Questionnaires from Tourists Leaving the Monkey Forests at Pandangtegal and Sangeh**[1]

		Both Sites			**Padangtegal**	**Sangeh**	
Variable description:	***N***	**Yes%**	**No%**	**t-test**	**Yes%**	**Yes%**	**X**
1) Did you come for the monkeys?	121	81	19	**	76	97	$p<.050$
2) Did you feed the monkeys?	123	48	52	NS	36	82	$p<.001$
3) Are you afraid of monkeys?	115	42	58	NS	35	63	$p<.050$
4) Did the monkeys climb on you?	123	40	60	**	21	91	$p<.001$
5) If yes, did you like the monkeys climbing on you?	48	44	56	NS	42	45	NS
6) Did they take anything?	121	14	86	**	3	44	$p<.001$
7) Did they bite you?	123	6	94	**	2	16	$p<.050$
8) Should tourists feed monkeys?	113	39	61	**	35	52	NS
9) Should local people feed them for you?	103	66	34	**	63	77	NS
10) Did you read the forest signs?	89	80	20	**	—	—	—
11) Would you recommend a visit?	111	87	13	**	85	92	NS

[1] Key: *N* = sample size; t-tests are between group percents; values of p are for two-tailed tests; NS = not significant; X = chi-square between sites; * = $p<.05$; ** = $p<.001$.

Table 4.3. **The Results of Following and Observing Tourists Who Brought Food into the Monkey Forests at Padangtegal and Sangeh**[1]

Variable Number and Description	*N*	Both Sites Yes%	Both Sites No%	T-test	Ubud Yes%	Sangeh Yes%	Chi-square between Sites
1) Was the sample in the morning (= yes) or in the evening (=no)?	39	49	51	NS	50	40	NS
2) What was the sample duration (min.)? Yes = Mean; No = Median	35	6	6		5	10	p<.01
3) How many monkeys were in the sample? Yes = Mean	39	7			6	10	p<.01
4) What was the food type fed to the monkeys?							
a) peanut	15	52		—	42	67	
b) banana	13	45		—	46	17	
c) cake	1	3		—	0	17	
5) Did a monkey refuse food during the sample?	39	41	59	NS	55	0	p<.01
6) Did a monkey bite a tourist during the sample?	39	18	82	p<.001	10	40	NS
7) Did a monkey hit a tourist or guide during the sample?	39	36	64	p<.03	41	20	NS
8) Did a monkey steal anything from a tourist other than food?	39	15	85	p<.001	10	30	NS
9) Did a monkey climb on or touch a tourist?	39	77	23	p<.001	69	100	NS
10) Was there monkey aggression on a tourist?	39	87	13	p<.001	83	100	NS
11) Was aggression rewarded?	39	80	20	p<.001	76	90	NS
12) Was non-aggression rewarded?	39	56	44	NS	66	30	NS
13) Was the reward for aggression a bag of peanuts or a bunch of bananas?	39	21	79	NS	24	10	NS
14) Was the reward for non-aggression a bag of peanuts or a bunch of bananas?	39	0	100	p<.01	0	0	—
15) Was there monkey aggression on monkey?	39	49	51	NS	48	50	NS
16) Did a tourist hit a monkey during the sample?	39	3	97	p<.001	0	10	NS
17) Did a tourist or guide defend the food with a rock, stick, or slingshot?	38	18	82	p<.001	7	56	p<.01
18) What was the nationality of the tourist?							
a) Australian/New Zealander	7	22			27	—	—
b) European	12	37			38	33	—
c) American/Canadian	1	3			4	—	—
d) Japanese	7	22			23	17	—
e) Indonesian	5	16			8	50	—

The most interesting result in following the tourists is that they appear to be training monkeys to be aggressive. In other words, monkeys who attack tourists are rewarded with food. At Padangtegal, tourists entered the Monkey Forest and bought peanuts or bananas at the front gate near the Feeding Station (see the map on page 114). As they walked down the Gate Road toward the Garuda Area, hungry monkeys often besieged them (see Plate 4.4). A frightened tourist readily gives up food. Aggression is positively reinforced by food rewards. Not only that, but the intensity of aggression corresponds with the quality and quantity of tourist food. We scored aggression as food rewarded when there was a close temporal (one or two seconds) and, to us, causal connection between monkey aggression on a tourist and the receipt of food from that tourist by the monkey. The statistically significant positive correlations between tourist feeding and monkey aggression suggest that operant conditioning of aggression has taken place. At Padangtegal and Sangeh, the amount of bananas at the former site and peanuts at the latter site was significantly positively correlated with food-rewarded aggression. The more bananas a tourist had, for example, the more aggressive and the faster the monkeys were in getting those bananas. In fact, monkeys were never rewarded with lots of food when they were not aggressive. There was also a significant positive correlation between the frequency of redirected aggression and the frequency of all food. Monkeys attacked each other more when tourists had more food. At both sites, the tourists who fed monkeys were attacked significantly more than tourists who did not feed monkeys.

There was a big difference between the two sites. Sangeh had significantly higher frequencies of both food-related and non-food-related aggression than did Padangtegal. Monkeys at Sangeh attacked tourists regardless of whether or not the tourists had food. A major difference between the two monkey forests is that the guides at Sangeh are often photographers with Polaroids trying to earn extra money by taking pictures of tourists with monkeys on them. These guides then reward the monkeys for jumping on tourists by offering them food to jump off. The animals sometimes ripped earrings out of the ears of tourists or stole other items and ran into the forest. The animals were again rewarded or bribed by food in order for the tourists to retrieve these items. These events rarely happened at Pandangtegal. Guides or guards at Ubud did not carry cameras, and the managerial committee, not the tourists, paid them. The questionnaires also show that tourists at Sangeh were more afraid of monkeys than were tourists at Padangtegal. Tourists at the former site reported significantly more biting, theft, and climbing on them by monkeys than did tourists at the latter site. Perhaps another reason for the greater frequency of aggression by monkeys at Sangeh is that there were many more monkeys. I counted at least 400, and they appeared to be more underfed and desperate for food than the monkeys at Padangtegal. Animals at the latter site sometimes refused food, whereas at Sangeh the monkeys never refused food.

Some primatologists might argue that such behavior by monkeys is "unnatural." Climbing on tourists or pulling down skirts is surely a novel way to get food, unlikely to be in the behavioral repertoire of monkeys deep in the

Plate 4.4. *This tourist in the Monkey Forest is about to relinquish his bananas after two monkeys climb on him.*

rain forests of Borneo. However, it is a learned, adaptive behavior for animals in a densely populated monkey forest. Raiding the farmers' fields can result in injuries from the sickles of angry farmers. Tourists are a lot less dangerous than farmers. Tourists sometimes got sticks or used cigarette lighters (the flame thrower defense) or simply played keep away with the food, tossing it back and forth with an animal running after it. Monkeys recognized some farmers when they walked through the forest. The animals would give alarm calls and flee from such individuals, who were known by us to aggressively chase monkeys in the rice fields.

Most of the monkey bites on tourists that I witnessed occurred when monkeys protected their babies and/or when tourists unreasonably provoked animals. Tourists, for example, would sometimes grab infants, causing them to scream. The reaction of other animals was immediate. They would rush to the aid of the infant, threaten, and sometimes bite the tourist until the infant was released or escaped. The most serious bite I ever saw occurred when a tourist went within a meter of a dying infant, ignoring the threats of nearby monkeys, and was bitten on the arm while taking photos. Despite these rare occurrences, we were actually quite impressed with the restraint some of these monkeys exhibited. For example, some tourists would sit on the road in the middle of the Monkey Forest and offer all kinds of fruits to the animals. I have seen such tourists deny these prized fruits, saying no, no, no to adult males, who would then open their mouths with their canines a few feet away from the necks of these sitting tourists. Despite this teasing, the animals did not bite. In all my years of watching monkeys, I have never been bitten.

The wounds caused by animal bites are not the only problem. Monkey bites are rarely fatal, but Sheo Singh (1969) has reported a fatality in India.

Monkeys can also cause other serious problems. In China, Q. K. Zhao (1994) reported that falling while fleeing from monkeys caused at least ten tourist deaths. Transmission of diseases to humans is also a concern. Such diseases as herpes B, hepatitis A, dengue fever, yellow fever, enterovirus, coronavirus, encephalomyocarditis, monkey pox, Marburg's disease, shigella, salmonella, and many parasites are all associated with monkey-human interactions (Eduardo & Castro, 1988; Kalter & Heberling, 1992; King, Yarbrough, & Anderson, 1988; Lapin & Shevtsova, 1992). In 1994, streptococcus B haemolytic killed more than a hundred monkeys at three different monkey forests in Bali (Harya Putra, pers. comm.). Hundreds of pigs also died from the disease, and it is believed that they transmitted it to the monkeys. Tourists worried about getting diseases from monkeys, especially rabies, but we are unaware of any such transmission at any of the Balinese monkey forests. Bali is said to be free from rabies (Mabbett, 1985).

Although our samples are small, some of the other interesting results in Table 4.2 are that most of the tourists went there specifically to see the monkeys and that they enjoyed their visit. Most of the tourists questioned said they would recommend it to a friend. Of the 122 tourists responding to our question inquiring their nationality, over 50 percent said they were from Europe, 22 percent were from Australia or New Zealand, and 16 percent were from North America. Many more Indonesian tourists went to Sangeh than to Padangetegal.

There is a noticeable managerial difference between the monkey forests of Padangtegal and Sangeh. Some of the more obvious differences are that Padangtegal uses numbered tickets, unlike Sangeh, and that warning signs and other signs for visitor safety, in addition to maps, were prominently displayed at Pandangtegal. This is despite that fact that Sangeh was more dangerous. Forty percent of all tourists at Sangeh were bitten as opposed to 10 percent at Padangtegal. We discussed these questionnaire results with the managerial committee and other people of Padangtegal. We suggested that signs in languages other than English, such as German or Japanese, be considered. The committee said that they would train guides in various languages and that the guides, rather than the tourists, would feed the monkeys while tourists were present. Our questionnaires also showed that most of the tourists felt the same way, especially after seeing what monkeys can do. One of the most important things management could do is to end the food-rewarded aggression. Padangtegal was already taking steps to do this. We took several observation samples of guards while they fed monkeys. None of the aggression exhibited by the animals was rewarded, in fact, just the reverse. The guards rewarded nonaggression. There was not only significantly less aggression on guards but also significantly less aggression by monkeys on other monkeys (redirected aggression) while the guards fed the animals as compared to the samples of tourist feeding.

In 1995, together with my Indonesian colleague, Harya Putra, we suggested a hypothesis accounting for the different management styles at the two monkey forests (Wheatley & Harya Putra, 1995). The tremendous growth in tourism in Ubud precipitated a restructuring of community relationships along the traditional Balinese Hindu philosophy of Tri Hita Karana. The great increase in wealth has the power to corrupt and to destroy community life, but by giving this wealth to God, they can balance its corrupting influence and spread the wealth equitably. Sangeh is off the regular tourist track and does not appear to be besieged by tourists, Western or otherwise. The desa adat at Sangeh does not perceive the more immediate threat of tourism, and its management style has not been reformed. We therefore concluded that the strong conservation program developed at Wenara Wana in Padangtegal was only one part of the three-part relationship that people have with nature and that the core of this reorganization was a reassertation of their Balinese Hindu religion. The ability to balance the new problems of ecotourism with the new revenue sources that ecotourism provides allows this community to restore the three harmonious relationships of the doctrine of Tri Hita Karana. Consequently, Padangtegal's recent and apparently unique success in benefiting from ecotourism may serve as a useful model for conservation and other purposes elsewhere.

Chapter 5

Cultural Primatology

Anthropologists work in a unique and exciting field. Our emphases on fieldwork and languages give us firsthand knowledge of a culture. Our training in understanding a foreign culture also includes such concepts as our holistic approach and our utilization of cultural relativism. The holistic approach explores *all* aspects of a culture to properly understand it. This should include the culture's views toward nature and animals. Cultural relativism is the concept that those aspects of a culture are best understood within that culture's worldview. Furthermore, by including primatology within its borders, anthropology recognizes the contribution that the former can make in understanding ourselves, which is, after all, the goal of anthropology. Perhaps a better way of explaining the above is to say that other disciplines may not provide their practitioners with an adequate understanding of another culture, its language, and an appreciation of the culture's long history and prehistory with their object of study. For example, a nonanthropologist might view the fieldwork experience with a nonhuman primate as an experimenter in a laboratory with the minor inconvenience of having to deal with strange people and strange diseases. Perhaps too, an unwritten attitude is that the only information worth getting is from animals who have no human interactions at all. As if that were even possible!

One goal of this book and of this research is to bridge the gulf between the subfields in the discipline of anthropology. On the one hand, cultural anthropologists focus on one species and one subspecies, *Homo sapiens sapiens*, at more or less in one point in time: the present. On the other hand, primatologists within the subfield of physical anthropology are those researchers who study several hundred species of primates with the exception of us, a fellow primate. The gulf between these areas is not limited to the field of anthropology. Many primatologists are not anthropologists. Their fields are in such disciplines as psychology, biology, or ecology. Such diversity of fields certainly gives primatological research greater breadth and depth, but the tendency to exclude humans and human influences may be even greater in other fields than in anthropology. Although some habitats and some species are more affected by humans than others, a world without humans just does not exist. No ecosystem on the earth is untouched by humans.

The area of cultural primatology investigates the interactions between human and nonhuman primates. It is a symbiotic area between the subfields and certainly within the holistic tradition of anthropology. It examines all facets of this interaction between humans and nonhuman primates. While the holistic approach is often cited as the cornerstone and core of our discipline, in practice, an anthropologist's approach is often confined within a single subfield.

Cultural primatology, therefore, goes beyond this, by viewing the holistic approach as a multisubfield one.

The term *cultural primatology* has been used before, but with another meaning. Richard Wrangham, F. De Waal, and W. C. McGrew (1994) used it to refer to the primatological investigation of culture. While such an investigation is a good example of the synergism between the subfields, it is not as broad as the meaning that I intend. Leslie Sponsel (1997) has proposed another term, ethnoprimatology, for cultural primatology. He defines ethnoprimatology as an "interface between human and primate ecology." As an example of the lack of research in this area, he points out, that in a search of the primatological literature, he did not find a single article that included humans and other primates together in the same ecosystem. Sponsel's outline of this field appears to be a little more ecologically oriented but very close to the broader meaning that I have for cultural primatology. Whatever term is used, it is important to reiterate the importance of human influences on nonhuman primates and vice versa.

What are the benefits of utilizing cultural primatology and taking a broader holistic view? How are we better off for it? Beginning with the relevance for cultural anthropologists, I think the first chapter of this book is a good example of what the subfield of cultural anthropology can learn from the cultural primatological approach. Monkeys are more than just another animal. The name *Bali* may derive from the name of a monkey king mentioned in the Ramayana Kakawin. Such a possibility has either been overlooked or ignored—or perhaps even subconsciously denied—by countless other authors of books on Bali. As the first chapter points out, many authors state that Bali means "offering," but no one explains why Bali is an offering. The stories of Bali, the monkey king, however, do explain it. The validity of my proposal is also given added weight through other lines of investigation, for example, through other manuscripts, as I mention.

Another benefit that cultural primatology makes apparent is in the reflexive nature of our discipline. What can we learn about ourselves and other primates by studying the Balinese? The second half of the first chapter of this book makes this explicit. Westerners have a long tradition of viewing nonhuman primates as loathsome and disgusting animals. Even Karl Marx (1853) used the monkey metaphor in his ethnocentric justification of English colonialism in Asia when he stated that the "adoration of Hanuman, the monkey, and Sabbala, the cow" was proof those Hindu communities and other nature worshippers had degraded man rather than elevated him. We, therefore, tend to overlook the other side of monkeys or tend to be less than impartial on their real nature (see Plate 5.1). We easily see the Balinese aversion to animality but

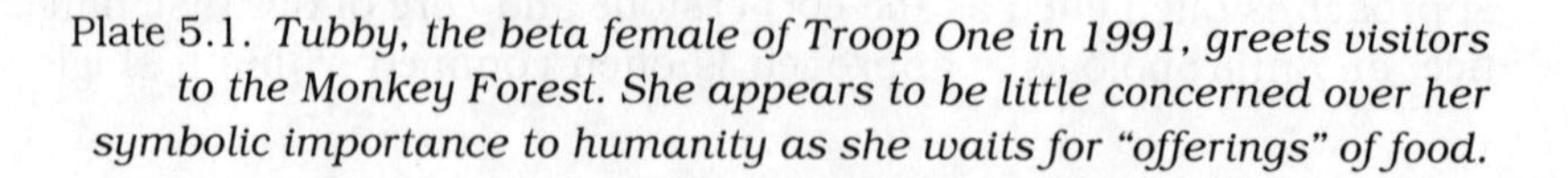

Plate 5.1. *Tubby, the beta female of Troop One in 1991, greets visitors to the Monkey Forest. She appears to be little concerned over her symbolic importance to humanity as she waits for "offerings" of food.*

quickly dismiss or overlook the possibility of monkeys as demigods, much less incorporate the possibility that monkeys can be both good and evil.

A further benefit of the approach of cultural primatology is to show primatologists how important humans have been in the dispersal and in the adaptations of this species of macaque. The antiquity of primate commensalism is little realized and little appreciated. Calling such commensalism "unnatural" misses the point: it is natural. They have been with us for a long time. The long-tailed macaque as well as other macaques have been called "weed" species—a pejorative term reflecting their ubiquity and undesirability. They certainly can cause havoc in our gardens and in our houses, but there is also a positive side. It should be obvious from this book that the interactions between monkeys and Balinese people are commensal. Both species benefit. An understanding of human-animal relationships is the first step toward conservation.

While many primatologists might not consider studying these macaques under such conditions, I welcomed the opportunity. My research, over a seven-year period of time, is the first long-term research on this species of macaque. I can only hope that my work will provide as useful a guide to further research on this species as it was in other long-term studies of other species of provisioned macaques. The information from the Monkey Forest at Padangtegal on such aspects as female behavior and intra- and intertroop competition will be difficult to duplicate, at best, on unhabituated troops.

Last, and perhaps most importantly, the benefit of cultural primatology can be seen in the area of conservation. Anthropologists can play an important role in applying their understanding of commensal relationships toward the conservation of nonhuman primates (Orlove & Brush, 1996). Bali is neither unusual in its worldview nor in the threat to its wildlife and livelihood from the outside world. For some examples, see M. S. Hood (1993). Such destruction makes cultural primatology that much more urgent because more than just nonhuman primates are at stake. It is precisely in this area of the application of anthropology to world problems that our holistic method is showcased. This is exactly what both Annette Weiner (1995) and James Peacock (1997) advocate to resurrect our discipline's multisubfield approach. In fact, there is irony in the imminent breakup of our field when our newest subfield, applied anthropology, is becoming larger and more popular than ever before. There are many examples of applied anthropology, our discipline's fifth subfield. All anthropologists know, for example, about the problems of the Green Revolution. The Green Revolution used new agricultural technologies and bureaucracies in an attempt to help solve world hunger. How the Green Revolution affected Bali is also well-known (Lansing, 1991; Lansing & Kremer, 1993). The replacement of native varieties of rice with hybridized varieties, its continual cropping, and the required application of chemical fertilizers and pesticides led to disastrous crop losses. What I am suggesting here is that just as Western technology and governmental interference overlooked the role of local temple networks and traditional mechanisms of water distribution in Bali, so too, the conservation movement has been slow to realize the importance of local structures and beliefs in effecting conservation reform.

The Balinese Hindu philosophy of Tri Hita Karana is directly responsible for the conservation efforts at the Monkey Forest, according to the people of Padangtegal. This philosophy strikes me as extremely important, but I could find very little information in the anthropological literature on it. This philosophy appears to be crucial in the seemingly exceptional case of Ubud's adaptation to ecotourism and urbanization as well. If true, then the dire predictions of a "show-down between agriculture and tourism" (Stevens, 1994) may once again be premature just as many of the other dire predictions for the demise of Bali. It strikes me that it is exactly this process of knowing when a balance or imbalance has occurred among the three harmonious relationships as outlined in the Tri Hita Karana philosophy that is so important to the success of Bali. This constant community dialogue and constant reshaping of their world has allowed the Balinese to adapt to a continually changing world.

Appendix

The Vocal Repertoire of *M. fascicularis* at the Monkey Forest at Padangtegal

Sixteen different vocalizations were classified into nine categories based on previous descriptions of the calls of the genus *Macaca*. Spectrographic analysis of 750 calls were made, of which 616 are indicated below. The patterns of these calls are described below. Precise determination of minimum frequencies for the threats and grunts was difficult because of the background noise.

1) *Kra call.* The various characteristics used by Ryne Palombit (1992a) in defining the different types of kra calls were analyzed. The results are presented in table A1. Table A1 shows that 53 percent of the calls had brief first pulses. The mean frequency of the first pulse was 5.84 kHz with a range of 2–16 kHz. Only 5 percent of our calls had a first pulse frequency of 9 kHz or more. Almost all of our calls (96 percent) had *most* of their energy between 3 and 8 kHz although some had energy over 8 kHz. Most of the calls (77 percent) had separation between the pulses. The number of pulses ranged from one to six. The presence of reverberation at the end of the call was variable even at close range and did not depend on the distance from the caller. Tonal sweeps were present in 31 percent of the calls, and these tended to occur in the first few pulses. The first pulse sometimes had tonal upsweeps or downsweeps, an important element in *M. fuscata* (Masataka, 1983). The context of most of our calls was difficult to identify, as many of them were directed toward a pile of rocks. It is apparent that most of our calls have overlap in the variables used by Palombit (1992a) in differentiating kra-c calls from kra-a calls in Bornean *M. fascicularis*. For example, one of our calls had a brief first pulse with most of its energy between 3 and 8 kHz, but the frequency of the first pulse was 10.5 kHz, and its maximum frequency was 12. Most of our calls would be kra-c if we used the variables of brief first pulse, most call energy between 3 and 8 kHz, and low first pulse frequency. Using the variable of sharp separation between pulses and maximum frequency, however, most of our calls would be kra-a. There was no significant difference in the maximum frequency when the first pulses were brief or not. The mean maximum frequency was 8.01 kHz when the first pulse was brief, but the mean maximum frequency was 7.43 kHz when the first pulse was not brief. Since we know most of the individuals giving the call, we can also point to variation within variables. For example, there were two adult females who seemed to call during a similar context. One caller had a continuum of maximum frequencies ranging from 5 to 11 kHz with pulses ranging from 2 to 6, and the other caller had a continuum of maximum frequencies ranging from 6 to 12 kHz with pulses ranging from 1 to 3. Another kra call was given to a farmer's dog. It was soft and high-pitched

and is very similar to the alarms to dogs and humans as described by J. O. Caldecott (1986) and to the alarm to minor mammalian predators depicted by Dorothy Cheney and Robert Seyfarth (1990).

Table A1. **Acoustic Parameters of the Kra Call (*N* = 109; minimum number of individuals = 9; maximum number of individuals = 16)**

		Frequency	%	Mean	Standard Deviation
Tonal sweep	yes	34	31.2		
	no	75	68.8		
Brief first pulse	yes	58	53.2		
	no	51	46.8		
Most call energy 3–8 kHz	yes	105	96.3		
	no	4	3.7		
Separation between pulses	yes	84	77.1		
	no	25	22.9		
Reverberation at end	yes	87	79.8		
	no	22	20.2		
Frequency first pulse				5.84	2.18
Minimum frequency				.92	.19
Maximum frequency				7.74	2.99
Call duration (milliseconds)				206.62	64.13
Number of pulses				2.72	1.04

2) *Threats*. After examining the sonograms, we noticed three types:
   a) *Threat rattle*. The calls have rapid pulsing or narrow segmentation usually with one band of high energy at low frequency (1 kHz) and usually a fainter band at about 3–4 kHz (*N* = 29) (See Figure A1e). Grunts generally lack the band of high energy at low frequency. The maximum frequency ranges between 4 and 11 kHz with an average maximum frequency of 6.5 kHz. Durations range from .15 to .75 seconds, with an average of .35 seconds. The minimum frequencies are either 500 or 1,000 kHz.
   b) *Bark kra*. These are temporally segmented or pulsed calls as described by Palombit (1992a). (See Table A2 and Figure A1d). The maximum frequency ranges from about 5 kHz to about 10 kHz (*N* = 7). The average maximum frequency is 8 kHz. The minimum frequencies of bark kras were slightly below 500 Hz.

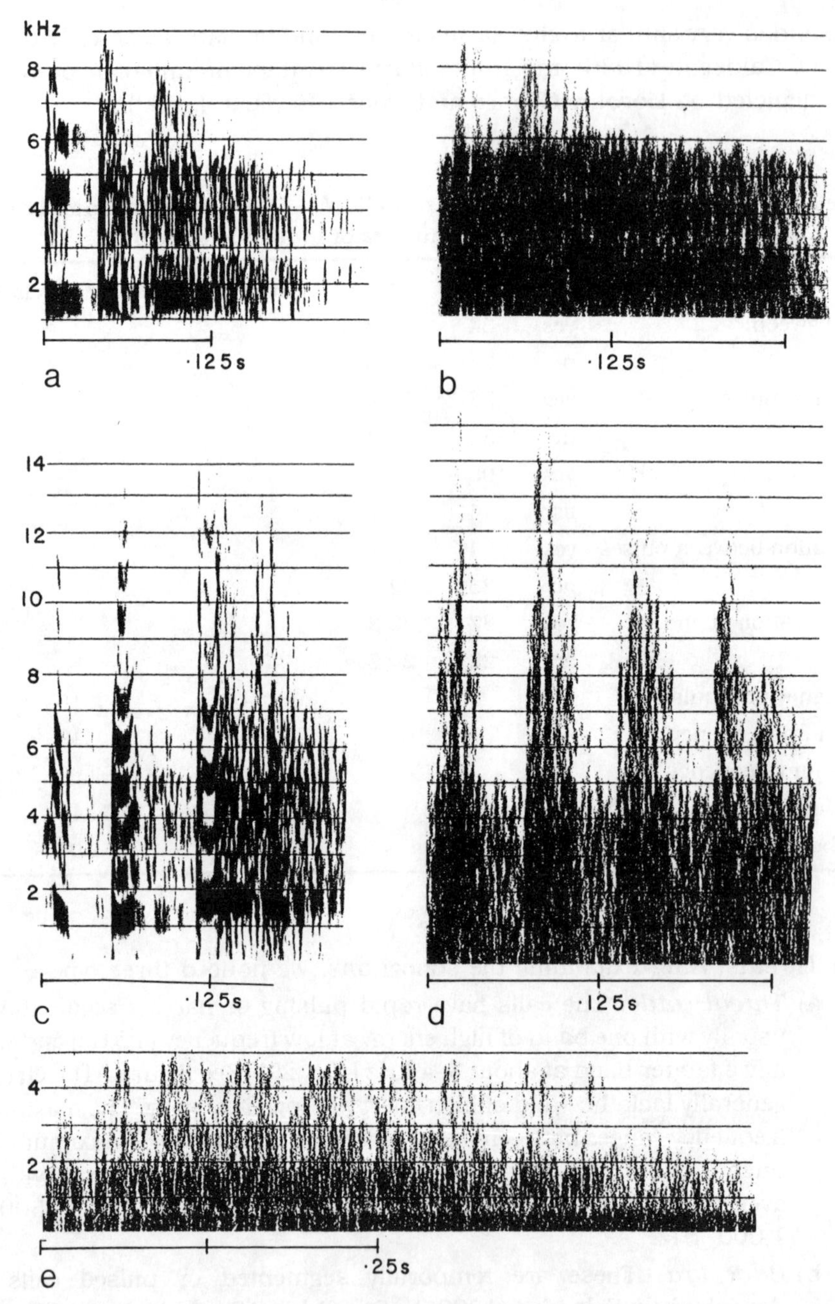

Figure A1. *Spectrograms: (a) common kra given by an adult female to a "pile of rocks"; (b) bark given by the alpha male; (c) aroused kra given by a young juvenile female to a "pile of rocks"; (d) bark kra given during an attack by an adult male on a young juvenile involving an adult female; and (e) threat rattle by the alpha male. The X-axis indicates time in tenths of a second, and the Y-axis indicates frequency in units of 2 kHz (2,000 cycles/second).*

Table A2. **Acoustic Parameters of the Bark Kra (N = 7)**

	Minimum	Maximum	Mean	Standard Deviation
Maximum frequency	4.4	10	5.77	1.923
Duration (milliseconds)	.25	.68	.398	.148
Number of Pulses	4	9	6.29	1.80

c) *Bark*. The pulses are fused in this call (See Figure A1b). The maximum frequency ranges from 7–8 kHz, and the average maximum frequency is 7.75 kHz (N = 4). The minimum frequencies are just under 800 Hz.

3) *Krahoo*. A low, loud, continuous two-syllable vocalization usually given by adult males (see Table A3 and Figure A2b). The frequency ranges between 2–8 kHz from at least three adult males in three different troops (N = 22). The maximum frequency averages 5.6 kHz and ranges from 4–8 kHz. It appears to be similar to baboons (Byrne, 1981; Kudo, 1987). It is a rapid and forceful exhalation followed by a rapid and forceful inhalation. The initial harsh segment to the call is not similar to the kra call in contrast to the call described by Palombit (1992a). Palombit (1992a) describes an inspirated bark called the bark-huh, but this bark is at lower amplitude than the krahoo.

Table A3. **Acoustic Parameters of the Krahoo (N = 18; 3 adult males)**

	Minimum	Maximum	Mean	Standard Deviation
Maximum frequency	4	6	4.94	.8024
Duration (milliseconds)	161	625	444.28	136.8

4) *Screams*. These were uninterrupted or continual vocalizations as described by S. Gouzoules, H. Gouzoules, and P. Marler (1984) for *M. mulatta* ; S. and H. Gouzoules (1989) for *M. nemestrina*. We identified five types of screams:

a) *Noisy scream*. This is an atonal noisy call of wide bandwidth given by "victims" of contact aggression (see Figure A1d). The maximum frequency ranges between 6 and 10 kHz, and the average maximum frequency is 7.5 kHz (N = 26). The minimum frequency ranges between 1–2 kHz. The duration ranges between .2 and .4 seconds and averages .3 seconds.

b) *Khreet screech*. This is an atonal call (partially tonal?) of sharply ascending and descending frequency modulations of one or more peaks (see Figure A2a). This call is similar to the call described by Palombit (1992a). Two variants are seen. The first variant is most similar to Palombit's (1992a) description. Only one sample of this variant had

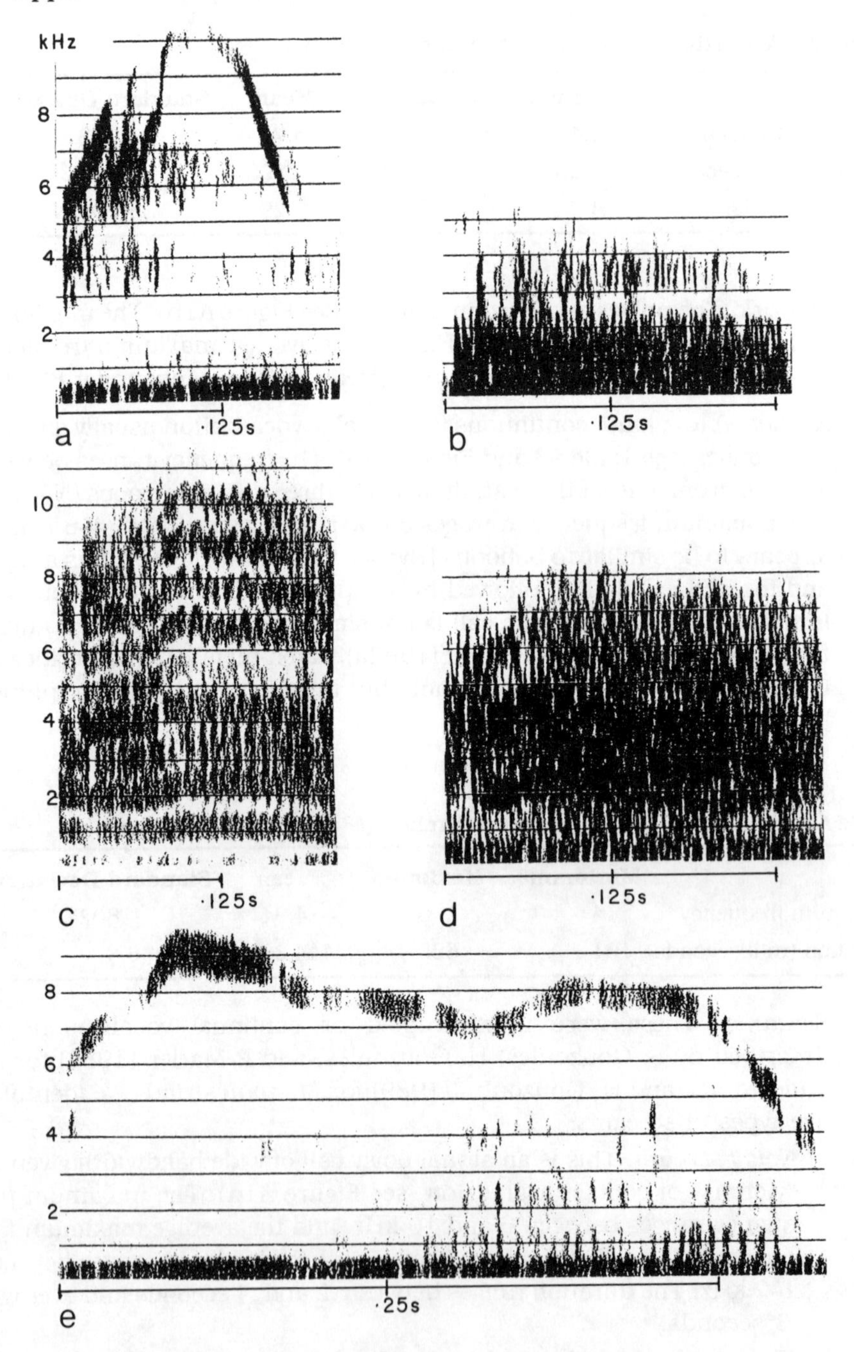

Figure A2. *Spectrograms: (a) khreet screech by a young juvenile who solicits (show-looks) two others to chase an older juvenile male; (b) wahoo by an adult male; (c) undulating scream to an older juvenile male; (d) noisy scream by an adult female after being bitten; and (e) tonal scream by an adult male who was then approached by infants.*

a peak. In the rest of our samples, the peak was absent. The maximum frequency ranges from 6 to 14 kHz, and its average is 9.0. The minimum frequency ranges from .5 to 3 kHz, and the duration ranges from .25 to .7 seconds (N = 22). The average duration is .4 seconds. The second variant has banded, tonal, and noisy elements (N = 45). The introduction frequently has noisy upsweeps, and the conclusion frequently has noisy downsweeps. The maximum frequency ranges from 6 to 11 kHz; the minimum frequency ranges from 1 to 3 kHz. One of these samples had an undulating scream at the end of a khreet screech. There were ten narrowly spaced harmonic resonances in this call. Its maximum frequency was 11 kHz, its minimum frequency was 1 kHz, and it was .65 seconds long. Only one undulating scream was recorded in isolation. Its maximum frequency was 12 kHz, its minimum frequency was 3 kHz, and it was .5 seconds long.

c) *Tonal scream*. This is an ascending and descending tonal call of long duration and narrow band widths (see Figure A2e). The maximum frequency ranged between 8 and 14 kHz, and its minimum frequency ranged between 1 and 5 kHz (N = 14). The average maximum frequency was 8.9 kHz and the average minimum frequency was 2.3 kHz. The duration ranged between .15 and 1.125 seconds, with an average duration of .52 seconds. Several calls had four tonal bands.

d) *Pulsed scream*. This is a pulsed call of short duration with 5–6 pulses per call (see Figure A3a). The maximum frequency ranges from 4 to 12 kHz, and it averages 9 kHz. (N = 9). The duration ranges from .15 to .475 seconds, and it averages .35 seconds. Tonal elements are usually present that distinguish this call from bark kras.

e) *Banded scream*. This long call appears to be a previously undescribed scream. Its name derives from its structure as it appears on the spectrogram (see Figure A3b). It usually begins with ascending sweeps with two main bands of energy that define the call's band width. There is sometimes a fainter third band between the other two bands. It is sometimes noisy and with modulations, but the bands persist. The maximum frequency ranges between 4 and 11 kHz. The average maximum frequency is 6.26 kHz (N = 83). The minimum frequency ranges from .5 to 4 kHz. The duration ranges from .13 to 1.5 seconds with an average of .63 seconds.

5) *Copulation call*. This is an atonal, plosive and rhythmic call given by females during copulation with most energy in the lower frequencies around 2 kHz (see Table A4 and Figure A3c). F. I. Deputte and M. Goustard (1980) have analyzed this call. Copulatory calls can be the longest of any call in the repertoire. One of our calls was 8.22 seconds long, with 52 pulses. The number of pulses or units in the call range from 3 to 52 (N = 20). The average number of pulses is 22. The maximum frequency ranges from 3 to 8 kHz and the average maximum frequency is 5.17 kHz. Tonal units were rare.

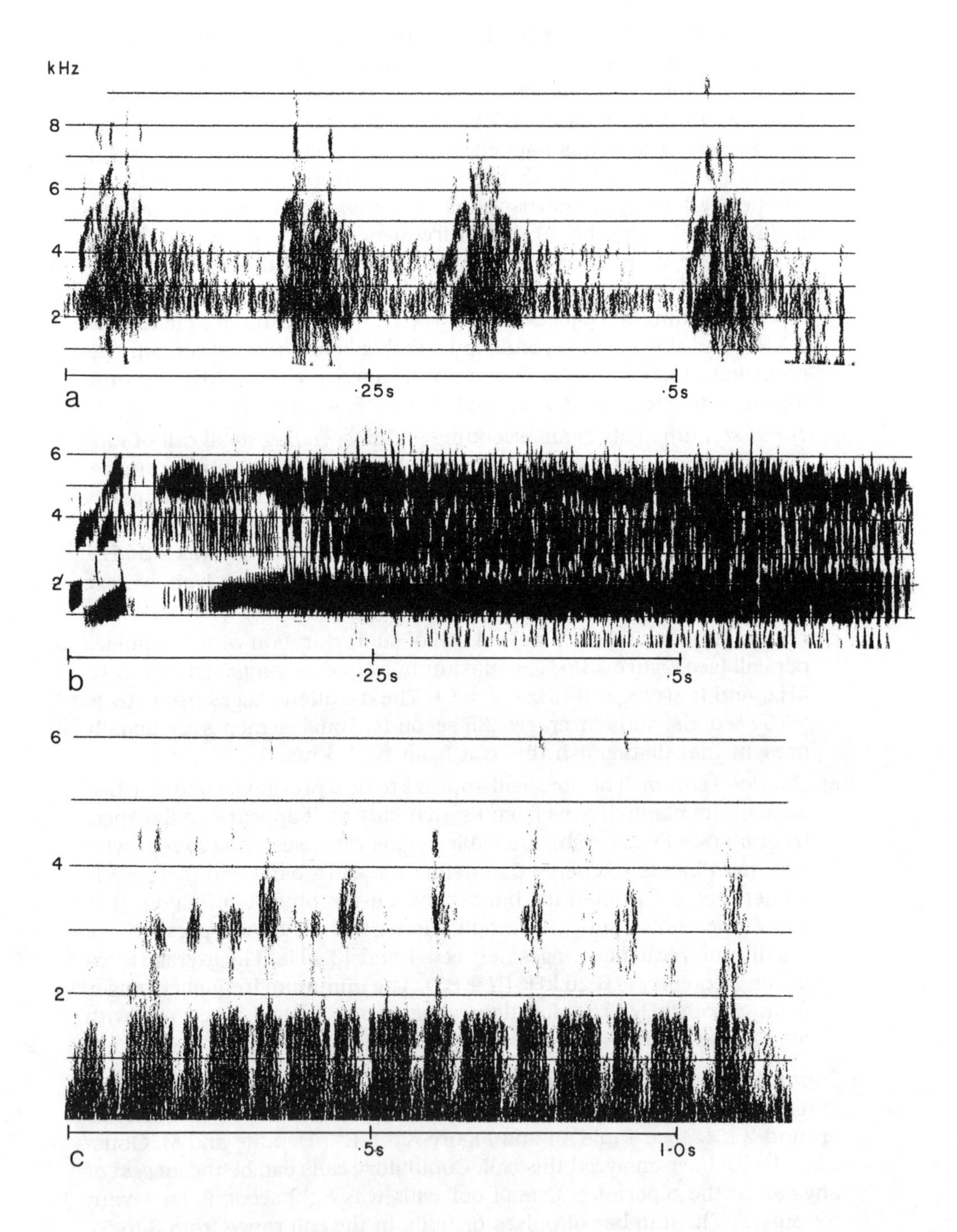

Figure A3. *Spectrograms: (a) pulsed scream by an adult female to an adult male; (b) banded scream by an adult female to an adult male; (c) copulation call by an adult female. Note the change in scale for the X and Y axes on the copulation call.*

Table A4. **Acoustic Parameters of the Copulatory Call (*N* = 20; 9 different adult females)**

	Minimum	Maximum	Mean	Standard Deviation
Maximum frequency	3	8	5.17	1.469
Duration (seconds)	1.43	8.22	.398	.148
Number of pulses	3	52	21.6	13.582

6) *Affiliation call.* This call is structurally similar to the call described by N. Masataka and B. Thierry (1993) for *M. tonkeana* (see Figure A4a). The maximum frequency ranged from 4 to 8 kHz with an average of 5.7 kHz. (*N* = 10). The durations range from about .075 to .2 seconds and average .12 seconds. The number of units per call ranged from 2 to 4.

7) *Contact or coo call.* We found two basic types of coo calls:

a) *Coo.* This is of variable duration, smooth, and of low frequency (up to about 5 kHz) with a great deal of variation as Steven Green (1975) found in *M. fuscata*. Our sonograms revealed many different types of coos (see Figure A4c). These calls seem to function in promoting friendly interactions and avoiding aggression, as has been previously described (Green, 1975; Palombit, 1992a). Most of our coos had no modulations unlike Palombit's (1992a) description. Masataka and Thierry (1993) mentioned that the more frequency modulated calls occurred as distances increased between the callers and the other troop members. The more constricted environment of Ubud may, therefore, explain this difference. Eighteen of our coos had no modulations or harmonics. Such calls had a maximum frequency ranging from 1 to 1.5 kHz and a minimum frequency ranging from .25 to 1.0 kHz. The durations ranged from .13 to .55 seconds with an average of .26 seconds. The contexts of these calls were subordinate adult females giving it to dominants, adults calling in isolation, and infants calling to their mothers. The calls with modulation and harmonics ranged from smooth or dipped early highs to smooth, late highs with 2 to 4 harmonics (*N* = 6). The calls with more harmonics had higher peak frequencies ranging up to 5.0 kHz. The various contexts included an adult female grooming a dead infant, an old infant trying to nurse, and infants separated from their mothers. Calls with harmonics ranged from .125 to .75 seconds, with a mean duration of .41 seconds. We also found coos with segmented bands, called pulsed whoo calls by Hohmann (1989) for *M. radiata*. The duration ranged from .8 to 1.8 seconds with intervals of about .1 to .25 seconds between the 3 to 4 segments (*N* = 4). The average duration was 1.5 seconds. These calls had 1–2 harmonics, and they occurred when a more dominant adult female held an infant of a low-ranking mother.

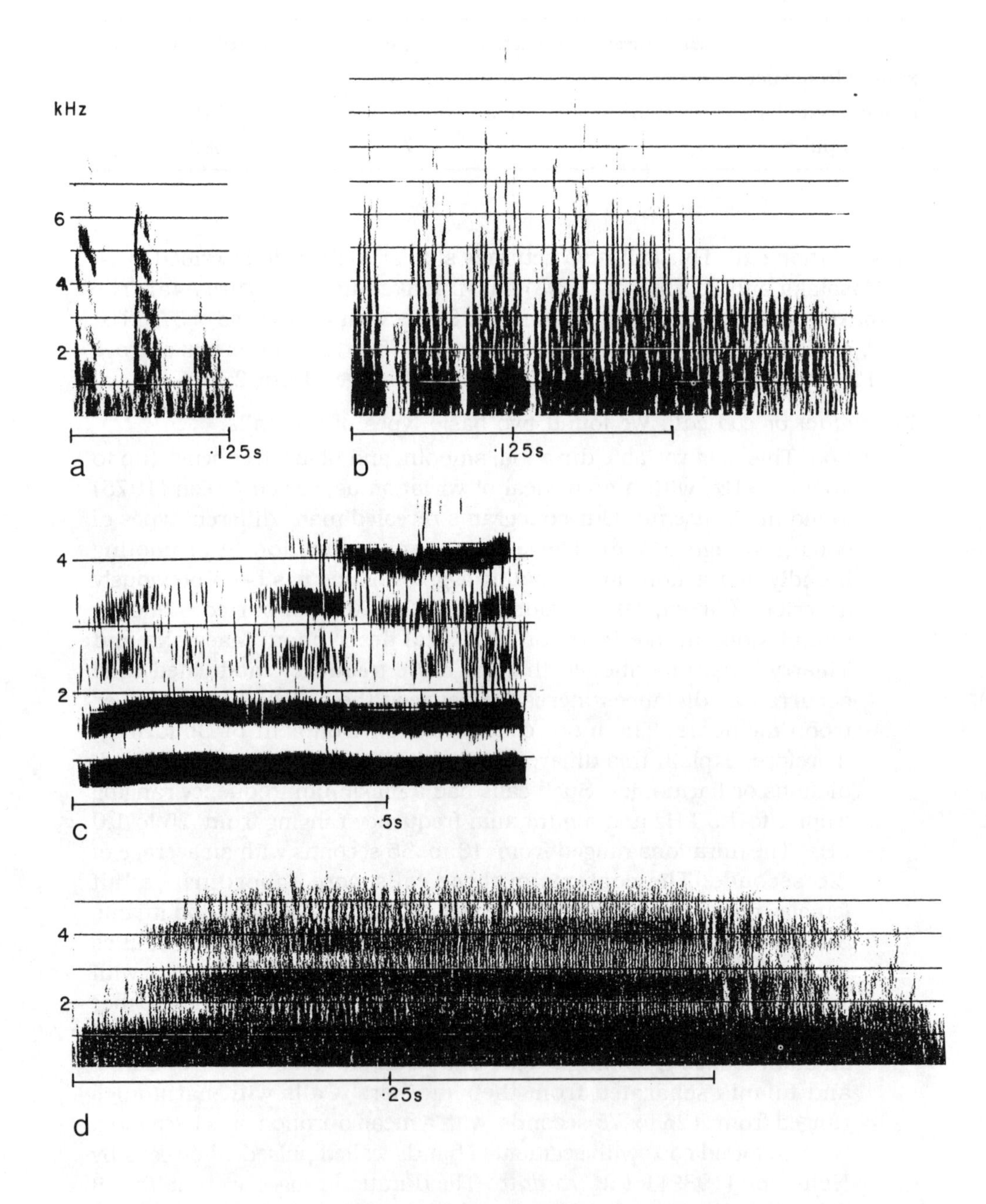

Figure A4. *Spectrograms: (a) affiliation call ; (b) grunt by an adult female who raises her eyebrows to another adult female (top right); (c) coo; and (d) grunt-coo. Note the change in scale for both the X and Y axes on the coo call.*

b) *Wraagh or grunt-coo*. Another type of coo had pulses similar to grunts (see Figure A4d). It usually had harmonics, with most of the energy at low frequency between 1 and 2 kHz. The maximum frequency ranged from 2 to 7 kHz, with an average of 4.6 kHz (N = 65). The duration ranged from .25 to .8 seconds, with an average of .46 seconds.

8) *Grunts*. This are calls of rapid, numerous, or staccato, pulses in one or several discrete units (see Figure A4b). They vary in frequency, duration, and spacing between pulses and spacing between pulse-units. The maximum frequency of the main energy ranges from 2 to 9 kHz. The average is 4.18 kHz (N = 150). Sometimes spikes of energy appear above the main energy pulses (N = 39). This call appears to have the most pulses for any call that we recorded. It is very similar to the grunts of *M. sinica* (Dittus, 1988), *M. arctoides* (Bertrand, 1969) and *Cercopithecus aethiops* (Cheney & Seyfarth, 1990).

9) *Geckers*. These calls are short, pulsed, and plosive. The maximum frequency was 7 kHz, and the gecker call occurred when an old infant was not allowed to nurse.

References

Aldrich-Blake, F.P.G. 1980. Long-tailed macaques. In *Malayan Forest Primates*. Ed. by D. J. Chivers. New York: Plenum Press. Pp. 147–165.
Alibasah, Margaret Muth. 1981. *Indonesian Folk Tales*. Jakarta, Indonesia: Djambatan.
———. 1990. *Folk Tales from Bali and Lombok*. Jakarta, Indonesia: Djambatan.
Angst, Walter. 1975. Basic data and concepts on the social organization of *Macaca fascicularis*. In *Primate Behavior: Developments in Field and Laboratory Research. Vol. 4*. Ed. by L. A. Rosenblum. New York: Academic Press. Pp. 325–388.
Ardika, I. Wayan, and Peter Bellwood. 1991. Sembiran: The beginnings of Indian contact with Bali. *Antiquity* 65: 221–232.
Ashton, P., and M. Ashton. 1972. *The Quarternary Era in Malesia*. Hull, UK: University of Hull, Department of Geography.
Asquith, Pamela J. 1991. Primate research groups in Japan: Orientations and East-West differences. In *The Monkeys of Arashiyama*. Ed. by L. M. Fedigan and P. J. Asquith. New York: State University of New York Press. Pp. 81–98.
Atmodjo, Sukarto. 1974. The Charter of Kapal. *Proceedings of the Sixth International Conference on Asian History*. Yogyakarta, Indonesia. Pp. 1–28.
Bagus, I Gusti Ngurah. 1968. *Arti dongeng Bali dalam pendidikan*. Singaradja, Indonesia: Direktorat Bahasa dan Kesusastraan.
Banks, Arthur S., and Thomas C. Muller, eds. 1998. *Political Handbook of the World*. Binghamton, NY: CSA.
Bateson, Gregory, and Margaret Mead. 1942. *Balinese Character: A Photographic Analysis*. New York: New York Academy of Sciences.
Beck, Benjamin. 1980. *Animal Tool Behavior: The Use and Manufacture of Tools by Animals*. New York: Garland Publications.
Bellwood, Peter. 1980. Plants, climate, and people: The early horticultural prehistory of Austronesia. In *Indonesia: The Making of a Culture*. Ed. by James J. Fox. Canberra, Australia: Research School of Pacific Studies, Australian National University. Pp. 55–74.
———. 1993. Cultural and biological differentiation in Peninsular Malaysia: The last 10,000 years. *Asian Perspectives* 32:37–60.
———. 1997. *Prehistory of the Indo-Malaysian Archipelago*. Honolulu: University of Hawaii Press.
Bertrand, M. 1969. The behavioural repertoire of the stumptail macaque: A descriptive and comparative study. *Biblio. Primatol.* 11:1–273.
Bhattacharji, Sukumari. 1970. *The Indian Theogony*. Cambridge: Cambridge University Press.
Bishop, N., S. Hrdy, J. Moore, and J. Teas. 1981. Qualitative and quantitative definitions of human influence in habitats of South Asian monkeys. *Int. J. Primatol.* 2(2): 153–167.
Blanford, W. T. 1887. Critical notes on the nomenclature of Indian mammals. *Proceedings of the Zoological Society of London*. Pp. 620–38.
Bruun, Ole, and Arne Kalland. 1992. *Asian perceptions of nature. Nordic Proceedings in Asian Studies No. 3*. Copenhagen: NIAS.

Budihardjo, Eko. 1986. *Architectural Conservation in Bali*. Gadjah Mada, Indonesia: Gadjah Mada University Press.

Buffon, Comte de. 1791. The nomenclature of apes. In *Climbing Man's Family Tree*. Ed. by Theodore D. McCown and Kenneth A. R. Kennedy, 1972. Englewood Cliffs, NJ: Prentice-Hall. Pp. 49–67.

Byrne, Richard W. 1981. Distance calls of Guinea baboons *(Papio papio)* in Senegal: An analysis of function. *Behaviour* 78:283–313.

Caldecott, J. O. 1986. An ecological and behavioral study of the pig-tailed macaque. In *Contributions to Primatology, Vol. 21*. Ed. by F. S. Szalay. Basel, Switzerland: Karger. Pp. 1–259.

Camille, Michael. 1992. *Image on the Edge: The Margins of Medieval Art*. Cambridge: Harvard University Press.

Chasen, N. C. 1940. A handlist of Malaysian mammals. *Bull. of the Raffles Museum, Straits Settlements No.15*:1–209.

Cheney, Dorothy L., and Robert M. Seyfarth. 1990. *How Monkeys See the World*. Chicago: University of Chicago Press.

Chiang, Mickey. 1967. Use of tools by wild macaque monkeys in Singapore. *Nature* 214:1258–1259.

Christensen, Hanne, and Ole Mertz. 1993. The risk avoidance strategy of traditional shifting cultivation in Borneo. *Sarawak Museum Journal* 44:1–18.

Covarrubias, Miguel. 1937. *Island of Bali*. Singapore: Oxford University Press. Reprinted 1986.

Creese, Helen. 1991. Balinese Babad as Historical Sources: A reinterpretation of the fall of Gelgel. *Bijdragen Tot de Taal-, Land-En Volkenkunde* 147:236–260.

Crockett, Carolyn, and W. Wilson. 1980. The ecological separation of *Macaca nemestrina* and *M. fascicularis* in Sumatra. In *The Macaques: Studies in Ecology, Behavior, and Evolution*. Ed. by D. Lindburg. New York: Van Nostrand Reinhold. Pp. 148–181.

Curtis, L. Perry Jr. 1997. *Apes and Angels: The Irishman in Victorian Caricature*. Rev. ed. Washington, DC: Smithsonian Institution Press.

Czech, Brian, and Paul Krausman. 1997. Distribution and causation of species endangerment in the United States. *Science* 277:1116.

Darlington, S. 1998. The ordination of a tree: The Buddhist ecology movement in Thailand. *Ethnology* 37(1): 1–15.

Davidson, Arnold I. 1991. The horror of monsters. In *The Boundaries of Humanity*. Ed. by James Sheehan and Morton Sosna. Berkeley and Los Angeles: University of California Press. Pp. 36–67.

De Josselin De Jong, P. E. 1983. Introduction: Structural anthropology in the Netherlands: Creature of circumstance. In *Structural Anthropology in the Netherlands*. Ed. by P. E. De Joss. Dordrecht, Holland: Foris Publications. Pp. 1–29.

De Waal, Frans B. M., Jan Van Hooff, and Willem Netto. 1976. An ethological analysis of types of agonistic interaction in a captive group of Java-monkeys *(Macaca fascicularis)*. *Primates* 17(3): 257–290.

De Zoete, Beryl, and Walter Spies. 1938. *Dance and Drama in Bali*. London: Faber & Faber.

Deputte, F. L., and M. Goustard. 1980. Copulatory vocalizations of female macaques (*Macaca fascicularis*) Variability factor analysis. *Primates* 21:83–99.

Dibia, I. Wayan. 1979. *Sinopsis tari Bali*. Sanggar Tari Bali "Waturenggong." Denpasar, Bali, Indonesia.

Dittus, Wolfgang P. J. 1988. An analysis of toque macaque cohesion calls from an ecological perspective. In *Primate Vocal Communication*. Ed. by D. Todt, P. Goedeking, and D. Symmes. Berlin: Springer-Verlag. Pp. 31–50.

Dobson, Andy, A. Bradshaw, and A. Baker. 1997. Hopes for the future: Restoration ecology and conservation biology. *Science* 277:515–522.

——, J. P. Rodriguez, W. M. Roberts, and D. S. Wilcove. 1997. Distribution and causation of species endangerment in the United States. *Science* 277:1117.

Dowson, John. 1972. *A Classical Dictionary of Hindu Mythology and Religion, Geography, History, and Literature*. London: Routledge and Kegan Paul.

Eduardo, S., and J. Castro. 1988. Some helminth parasites of the Philippine monkey *Macaca philippensis*. *Phil. J. Vet. Med.* 25(2): 15–21.

Eiseman, Fred B. Jr., 1989. *Bali: Sekala and Niskala. Vol. I.* Berkeley: Periplus Editions.

——, and Margaret Eiseman. 1988. *Woodcarvings of Bali*. Berkeley: Periplus Editions.

Emigh, John. 1984. Dealing with the demonic: Strategies for containment in Hindu iconography and performance. *Asian Theatre Journal* 1(1): 21–39.

Ermans, A. M., N. M. Mbulamoko, F. Delange, and R. Ahluwalia. 1980. *Role of Cassava in the Etiology of Endemic Goitre and Cretinism*. Ottawa, Canada: International Development Research Centre.

Eudey, Ardith. 1994. Temple and pet primates in Thailand. *Revue D'Ecologie* 49(3): 273–280.

Europa World Year Book. 1998. Vol. 1. London, UK: Europa Publications.

Farslow, Daniel. 1987. The behavior and ecology of the long-tailed macaque, *Macaca fascicularis* on Angaur Island, Palau, Micronesia. Ohio State University, unpublished Ph.D. diss.

Flores, Nona C. 1996. *Animals in the Middle Ages*. New York: Garland.

Fooden, Jack. 1964. Rhesus and crab-eating macaques: Intergradation in Thailand. *Science* 143:363–365.

——. 1991a. New perspectives on macaque evolution. In *Primatology Today*. Ed. by A. Ehara, T. Kimura, O. Takenaka, and T. Iwamoto. Amsterdam: Elsevier Science. Pp. 1–7.

——. 1991b. Systematic Review of Philippine Macaques. *Fieldiana Zoology N.S. No. 64*. Chicago: Field Museum of Natural History.

——. 1995. Systematic review of Southeast Asian longtail macaques, *Macaca fascicularis* (Raffles, 1821). *Fieldiana Zoology, N.S. No. 81*. Chicago: Field Museum of Natural History.

Forge, Anthony. 1978. *Balinese Traditional Painting*. Sydney: Australia Museum.

——. 1980. Tooth and fang in Bali. *Anthropology* 3:1–16.

Foster, Mary. 1979. Synthesis and antithesis in Balinese ritual. In *The Imagination of Reality: Essays in Southeast Asian Coherence Systems*. Ed. by A. L. Becker and A. A. Yengoyan. Norwood, NJ: Ablex Pub. Co. Pp. 175–196.

Franken, H. J. 1984. The festival of Jayaprana at Kalianget. In *Bali: Studies in Life, Thought and Ritual*. Ed. by J. L. Swellengrebel. Cinnaminson, NJ: Foris Publications. Pp. 235–265.

Frederick, William H., and Robert L. Worden, eds. 1993. *Indonesia, a Country Study*. Lanham, MD: Bernan Press.

Friederich, R. 1959. *The Civilization and Culture of Bali*. Calcutta: Susil Gupta (India) Private Ltd. Originally published in 1849–50.

Fugles, R. 1990. *Homer: The Iliad*. Viking Penguin.

Gadgill, M. 1985. Social restraints on resource utilization: The Indian experience. In *Culture and Conservation: The Human Dimension in Environment Planning*. Ed.

by J. McNeely and D. Pitt. London: Croom Helm. Pp. 135–154.
Galdikas, Birute, and Carey Yeager. 1984. Crocodile predation on a crab-eating macaque in Borneo. *Amer. J. of Primatol.* 6:49–51.
Gautier, J. P., and S. Biquand. 1994. Primate commensalism. *Revue D'Ecologie (Terre & Vie)* 49(3): 210–212.
Geertz, Clifford. 1980. *Negara: The Theatre State in Nineteenth Centùry Bali.* Princeton, NJ: Princeton University Press.
Geertz, Hildred. 1994. *Images of Power.* Honolulu: University of Hawaii Press.
———, and Clifford Geertz. 1975. *Kinship in Bali.* Chicago: University of Chicago Press.
Gibson, Kathleen. 1986. Cognition, brain size, and the extraction of embedded food resources. In *Primate Ontogeny, Cognition, and Social Behaviour.* Ed. by J. Else and P. Lee. Cambridge: Cambridge University Press. Pp. 93–103.
Glover, Ian C. 1971. Prehistoric research in Timor. In *Aboriginal Man and Environment in Australia.* Ed. by D. J. Mulvaney and J. Golson. Canberra: Australian National University Press. Pp. 158–181.
——. 1984. The late Stone Age in Eastern Indonesia. In *Prehistoric Indonesia.* Ed. by Pieter Van de Velde. Cinnaminson, NJ: Foris Publications. Pp. 273–295.
Gordon, D. J. 1943. The imagery of Ben Jonson's the masque of blacknesse and the masque of beautie. *Journal of the Warburg and Courtauld Institutes* 6:122–141.
Goris, Roelof. 1960. The religious character of the village community. In *Bali: Studies in Life, Thought and Ritual.* Ed. by J. L. Swellengrebel. Cinnaminson, NJ: Foris Publications. Pp. 77–100.
Goudrian, T., and C. Hooykaas. 1971. Stuti and stava of Balinese priests. *Verhandelingen der Koninklijke Nederlandse Akademie van Wetenschappen. Vol. 76.* London: North-Holland.
Gouzoules, S., and H. Gouzoules. 1989. Design features and developmental modification of pigtail macaque, *Macaca nemestrina*, agonistic screams. *Anim. Behav.* 37(3): 383–401.
———, and P. Marler. 1984. Rhesus monkey (*Macaca mulatta*) screams: Representational signalling in the recruitment of agonistic aid. *Anim. Behav.* 32:182–193.
Grader, C. J. 1990. The state temples of Mengwi. In *Structural Anthropology in the Netherlands.* Ed. by P. E. De Joss. Holland: Foris Publications. Pp. 155–188.
Gralapp, Leland W. 1967. Balinese painting and the wayang tradition. *Artibus Asiae* 29: 239–266.
Green, Steven. 1975. Variation of vocal pattern with social situation in the Japanese monkey (*Macaca fuscata*): A field study. In *Primate Behavior: Developments in Field and Laboratory Research, Vol. 2.* Ed. by L. A. Rosenblum. New York: Academic Press. Pp. 1–102.
Grove, R. 1992. Origins of western environmentalism. *Scientific American* 267(July): 42–47.
Hagerdal, Hans. 1995. Bali in the sixteenth and seventeenth centuries: Suggestions for a chronology of the Gelgel period. *Bijdragen tot de Taal-, Land- en Volkenkunde* 151(1): 101–124.
Harcourt, A. H. 1992. Coalitions and alliances: Are primates more complex than nonprimates? In *Coalitions and Alliances in Humans and Other Animals.* Ed. by A. H. Harcourt and F.B.M. De Waal. New York: Oxford University Press. Pp. 445–471.
Hauser, Marc D. 1993. The evolution of nonhuman primate vocalizations: Effects of phylogeny, body weight, and social context. *Amer. Nat.* 142(3): 528–542.
———. 1996. Vocal communication in macaques: Causes of variation. In *Evolution and Ecology of Macaque Societies.* Ed. by John E. Fa and Donald G. Lindburg.

Cambridge: Cambridge University Press. Pp. 551–577.

Henry, Patricia. 1987. The religion of balance. In *Indonesian Religions in Transition*. Ed. by R. Kipp and S. Rodgers. Tucson: University of Arizona Press. Pp. 98–112.

Hill, W.C.O. 1974. Primates. *Comparative Anatomy and Taxonomy. VII. Cynopithecinae*. New York: John Wiley & Sons.

Hinzler, Hedi I. R. 1974. The Balinese babad. *Sixth International Conference on Asian History*. Yogyakarta, Indonesia. Pp. 1–19.

——. 1981. Bima swarga in Balinese wayang. *Verhandelingen van het Koninklijk Instituut voor taal-, Land-en Volkenkunde 90*. The Hague: Martinus Nijhoff.

——. 1983. The artist behind the drawings. *Indonesia Circle* 30:5–12.

——. 1990. Former kingdom of Mengwi. In *Bali, the Emerald Isle*. Ed. by Eric Oey. Lincolnwood, IL: Passport Books. P. 215.

——. 1993. Balinese palm-leaf manuscripts. *Bijdragen Tot de Taal-, Land-en Volkenkunde* 149:438–473.

Hobart, Angela, Urs Ramseyer, and Albert Leeman. 1996. *The Peoples of Bali*. Oxford: Blackwell Publishers.

Hobart, Mark. 1985. Is God Evil? In *The Anthropology of Evil*. Ed. by D. Parkin. Oxford: Basil Blackwell. Pp. 165–193.

——. 1990. The patience of plants: A note on agency in Bali. *Review of Indonesian and Malaysian Affairs* 24:90–135.

Hoelzer, G. A., and D. J. Melnick. 1996. Evolutionary relationships of the macaques. In *Evolution and Ecology of Macaque Societies*. Ed. by John E. Fa and Donald G. Lindburg. Cambridge: Cambridge University Press. Pp. 3–19.

Hohmann, G. 1989. Vocal communication of wild bonnet macaques (*Macaca radiata*). *Primates* 30(3): 325–345.

Hood, M. S. 1993. Man, forest and spirits: Images and survival among forest-dwellers of Malaysia. *Southeast Asian Studies* 30(4): 444–456.

Hoogerwerf, A. 1970. *Udjung Kulon*. Leiden: E. J. Brill.

Hooijer, Dirk A. 1952. Fossil mammals and the Plio-Pleistocene boundary in Java. *Koninklijke Nederlandse Akademie van Wetenschappen, Proceedings, Series B*. 55:436–443.

——. 1962. Prehistoric bone: The gibbons and monkeys of Niah Great Cave. *Sarawak Museum Journal N.S.* 19–20:428–449.

Hooykaas, Christiaan. 1958. *The Lay of Jaya Prana, the Balinese Uriah*. London: Luzac & Co.

——. 1964. *Agamatirtha Verhandelingen der Koninklijke Nederlandse akademie van wetenschappen afd. letterkunde Vol. 70, no. 4*. Amsterdam: N. V. Noord-Hollandesche Uitgevers Maalschappij.

——. 1970. *Kama and Kala*. Amsterdam: North-Holland Publishing.

——. 1974. *Cosmogeny and Creation in Balinese Tradition*. The Hague: Martinus Nijhoff.

——. 1978. *The Balinese Poem Basur: An Introduction to Magic*. The Hague: Martinus Nijhoff.

Hooykaas, Jacoba. 1961. The myth of the young cowherd and the little girl. *Bijdragen Tot de taal-, Land-en volkenkunde* 117(2): 267–278.

——. 1963. *Märchen aus Bali*. Zurich: Verlag Die Waage.

Howe, L.E.A. 1984. Gods, people, spirits and witches: The Balinese system of person definition. *Bijdragen tot de Taal-, Landen Volkenkunde* 140(2/3): 193–222.

——. 1989 Peace and violence in Bali: Culture and social organization. In *Societies at Peace: Anthropological Perspectives*. Ed. by Signe Howell and Roy Willis. New

York: Routledge. Pp. 100–116.

Hrdy, Sarah Blaffer. 1979. Infanticide among animals: A review, classification and examination of the implications for the reproductive strategies of females. *Ethol. Sociobiol.* 1:13–40.

Huffman, Michael. 1984. Stone-play of *Macaca fuscata* in Arashiyama B troop: Transmission of a non-adaptive behavior. *J. of Human Evolution* 13:725–35.

Hunter, T. M. 1988. Crime and punishment in Bali: Paintings from a Balinese hall of justice. *Review of Indonesian and Malaysian Affairs* 22(2): 62–113.

Husband, Timothy. 1980. *The Wild Man.* New York: Metropolitan Museum of Art.

Hussey, Antonia. 1989. Tourism in a Balinese village. *Geographical Review* 79(3): 311–325.

Jacobs, Julius. 1994. [1883] The Dutch vaccinator meets the king of Gianyar. In *Travelling to Bali.* Ed. by Adrian Vickers. Oxford: Oxford University Press. Pp. 64–71.

Janson, H. W. 1952. *Apes and Ape Lore in the Middle Ages and Renaissance.* London: Warburg Institute, University of London.

Jensen, Gordon D., and Luh Ketut Suryani. 1992. *The Balinese People.* Singapore: Oxford University Press.

Jessup, Helen I. 1990. *Court Arts of Indonesia.* New York: Asia Society Galleries.

Jewett, D. A., and W. R. Dukelow. 1972. Cyclicity and gestation length of *Macaca fascicularis. Primates* 13(3): 327–332.

Johns, Andrew, and Bettina G. Johns. 1995. Tropical forest primates and logging: Long-term coexistence? *Oxyx* 29(3): 205–211.

Jones, Russell. 1974. The early history of George Samuel Windsor Earl. *Proceedings of the Sixth International Conference on Asian History.* Yogyakarta, Indonesia. Pp. 1–11.

Kaler, Gusti Ketut. 1983. *Butir-butir Tercecer Tentang Adat.* Denpasar, Bali: Bali Agung.

Kalter, S., and R. Heberling. 1992. Viral infections in primate colonies and their detection. In *Topics in Primatology, Vol. 3.* Ed. by S. Matono, R. Tuttle, H. Ishida, and M. Goodman. Tokyo: University of Tokyo Press. Pp. 383–389.

Kawai, Masao, and H. Ohsawa. 1983. Ecology of Japanese monkeys, 1950–1959. Recent Progress of Natural Sciences in Japan. *Anthropology* 8:95–108.

Kawamoto, Yoshi, and Tb. M. Ischak. 1981. Genetic differentiation of the Indonesia crab-eating macaque (*Macaca fascicularis*). I. Preliminary report on blood protein polymorphism. *Primates* 22:237–52.

——, and Jatna Supriatna. 1984. Genetic variations within and between troops of the crab-eating macaque (*Macaca fascicularis*) on Sumatra, Java, Bali, Lombok, and Sumbawa, Indonesia. *Primates* 25:131–159.

——, K. Nozawa, and T. B. Ischak. 1981. Genetic variability and differentiation of local populations in the Indonesia crab-eating macaque (*Macaca fascicularis*). *Kyoto University Overseas Report of Studies on Indonesian Macaque* 1:15–39.

———, and B. Suryobroto. 1985. Gene constitution of crab-eating macaques (*Macaca fascicularis*) on Timor. *Kyoto University Overseas Research Report of Studies on Asian Non-human Primates* 4:35–40.

Kempers, A. J. Bernet. 1991. *Monumental Bali.* Singapore: Periplus Editions.

King, Fred, C. Yarbrough, and D. Anderson. 1988. Primates. *Science* 240:1475–1481.

Kinnaird, Margaret, and Timothy O'Brien. 1996. Ecotourism in the Tangkoko Dua Sudara Nature Reserve: Opening Pandora's box? *Oryx* 30(1): 65–73.

Kinzey, Warren. 1997. Synopsis of New World primates. In *New World Primates.* Ed. by W. Kinzey. New York: Aldine de Gruyter. Pp.169–324.

Kirch, P. V. 1997. *The Lapita Peoples.* Cambridge, MA: Blackwell.

Kitahara-Frisch, J. 1991. Culture and primatology: East and west. In *The Monkeys of Arashiyama*. Ed. by L. M. Fedigan and P. J. Asquith. New York: State University of New York Press. Pp. 74–80.

Koyama, N., A. Asnan, and N. Natsir. 1981. Socio-ecological study of the crab-eating monkeys in Indonesia. *Kyoto University Overseas Research Report of Studies on Indonesian Macaque. Kyoto University Primate Research Institute* 1:1–10.

Kroeber, A. L. 1928. Subhuman cultural beginnings. *Quarterly Review of Biology* 3:325–342.

Kudo, H. 1987. The study of vocal communication of wild mandrills in Cameroon in relation to their social structure. *Primates* 28(3): 289–308.

Kuiper, F.B.J. 1983. *Ancient Indian Cosmogony*. New Delhi: Vikas Pub. House.

Labang, D., and L. Medway. 1979. Preliminary assessments of the diversity and density of wild mammals, man and birds in alluvial forest in the Gunong Mulu National Park, Sarawak. In *The Abundance of Animals in Malesian Rain Forests*. Ed. by A. G. Marshall. Hull, UK: Department of Geography, University of Hull. Pp. 53–66.

Laksono, P. M. 1986. *Tradition in Javanese Social Structure, Kingdom and Countryside*. Gadjah Mada, Indonesia: Gadjah Mada University Press.

Lansing, J. Stephen. 1983. *The Three Worlds of Bali*. New York: Praeger.

——. 1991. *Priests and Programmers*. Princeton, NJ: Princeton University Press.

——, and James Kremer. 1993. Emergent properties of Balinese Water Temple networks: Coadaptation on a rugged fitness landscape. *American Anthropologist* 95(1): 97–114.

Lapin, B., and Z. Shevtsova. 1992. Simian viral infections in the Sukhumi monkey colony. In *Topics in Primatology, Vol 3*. Ed. by N. Itoigawa, Y. Sugiyama, G. Sackett, and R. Thompson. Tokyo: University of Tokyo Press. Pp. 417–424.

Last, Jef. 1994. After the revolution. In *Travelling to Bali*. Ed. by Adrian Vickers. Oxford: Oxford University Press. Pp. 118–127.

Lattin, Don. 1994. The trouble with Bali. *Utne Reader* May/June: 84–89.

Lovric, B.J.A. 1986. The art of healing and the craft of witches in a 'hot earth' village. *Review of Indonesian and Malaysian Affairs* 20(1): 68–99.

——. 1988. Balinese theatre: A metaphysics in action. *Asian Studies Association of Australia Review* 12(2): 35–45.

Lucas, P., and R. Corlett. 1991. Relationship between the diet of *Macaca fascicularis* and forest phenology. *Folia Primatol.* 57:201–215.

Mabbett, Hugh. 1985. *The Balinese*. Singapore: January Books.

Maestripieri, Dario. 1994. Social structure, infant handling, and mothering styles in group-living Old World monkeys. *Int. J. of Primatol.* 15(4): 531–553.

Malik, Igbal. 1988. Possibilities of self-sustenance of free ranging rhesus of Tughlaqabad. *Journal Bombay Nat. Hist. Society* 85(3): 578–584.

Marrison, G. E. 1986. Literary transmission in Bali. In *Cultural Contact and Textual Interpretation*. Ed. by C. D. Grijns and S. O. Robson. Cinnaminson, NJ: Foris Publications. Pp. 274–291.

Marsh, Clive W., and W. L. Wilson. 1981. *A Survey of Peninsular Malaysian Primates*. Kuala Lumpur, Malaysia: Universiti Kebangsaan Malaysia.

Marx, Karl. 1853. The British rule in India. In *Karl Marx on Colonialism and Modernization*. Ed. by Shlomo Avineri, 1968. New York: Doubleday. Pp. 83–89.

Masataka, N. 1983. Psycholingual analyses of alarm calls of Japanese monkeys (*Macaca fuscata*). *Am. J. Primatol.* 5:111–125.

——, and B. Thierry. 1993. Vocal communication of Tonkean macaques in confined environments. *Primates* 34(2): 169–180.

Mathews, Anna. 1994. The holy mountain erupts. In *Travelling to Bali*. Ed. by A. Vickers. Oxford: Oxford University Press. Pp. 128–140.

Matsubayashi, K., S. Gotoh, Y. Kawamoto, K. Nozawa, and J. Suzuki. 1989. Biological characteristics of crab-eating monkeys on Angaur Island. *Primate Research* 5:46–57.

McGrew, William C., and Carolyn Tutin. 1978. Evidence for a social custom in wild chimpanzees? *Man* 13:234–251.

McKean, Philip Frick. 1979. From purity to pollution? The Balinese ketjak (monkey dance) as symbolic form in transition. In *The Imagination of Reality: Essays in Southeast Asian Coherence Systems*. Ed. by A. L. Becker and A. A. Yengoyan. Norwood, NJ: Ablex. Pp. 293–302.

McNeely, Jeffrey A. 1992. Protected areas in a changing world: The management approaches that will be required to enable primates to survive into the 21st century. In *Topics in Primatology. Vol. 2*. Ed. by N. Itoigawa, Y. Sugiyama, G. Sackett, and R. Thompson. Tokyo: University of Tokyo Press. Pp. 373–383.

McPhee, Colin. 1948. Dance in Bali. *Dance Index* (7–8):156–207.

Medway, Lord. 1958. Food bone in Niah Cave excavations, (–1958). *Sarawak Museum Journal* 8:627–636.

——. 1964. Post-Pleistocene changes in the mammalian fauna of Borneo. *Studies in Speleology* 1(1): 33–37.

———. 1970. The monkeys of Sundaland. In *Old World Monkeys*. Ed. by J. R. Napier and P. H. Napier. New York: Academic Press. Pp. 513–553.

———. 1972. The quarternary mammals of Malesia: A review. In *The Quarternary Era in Malesia*. Ed. by P. Ashton and M. Ashton. Hull, UK: Department of Geography, University of Hull. Pp. 63–98.

Miller, D. B., and Jan Branson. 1989. Pollution in paradise: Hinduism and the subordination of women in Bali. In *Creating Indonesian Cultures*. Ed. by Paul Alexander. Sydney: University of Sydney, Oceania Publications. Pp. 91–112.

Miller, G. S., Jr. 1942. Zoological results of the George Vanderbilt Sumatran Expedition, 1936–1939. Part V. Mammals collected by Frederick. A.Ulmer, Jr., on Sumatra and Nias. *Proceedings of the Academy of Natural Sciences, Philadelphia* 94. Pp. 107–167.

Mittermier, Russell. 1987. Conservation of primates and their habitats. In *Primate Societies*. Ed. by B. Smuts, D. Cheney, R. Seyfarth, R. Wrangham, and T. Struhsaker. Chicago: University of Chicago Press. Pp. 477–490.

Moerdowo. 1958. *Reflections on Indonesian Arts and Culture*. Suyrabaya, Indonesia: Surabaya Publishing House.

Moertono, Soemarsaid. 1981. *State and Statecraft in Old Java: A Study of the Late Mataram Period, 16th to 19th Century*. Monograph Series, no. 43. Ithaca, NY: Cornell Modern Indonesia Project.

Mohr, E. 1945. Climate and soil in the Netherlands Indies. In *Science and Scientists in the Netherlands Indies*. Ed. by P. Honig and F. Verdoorn. New York: Board for the Netherlands, Surinam and Curacao. Pp. 250–254.

Morell, Virginia. 1993 Anthropology: Nature-culture battleground. *Science* 161:1798–1802.

Muninjaya, A. A. Gde. 1982. Balinese traditional healers in a changing world. *Annual Indonesian Lecture Series*. Pp. 35–41.

Napier, J. R., and P. H. Napier. 1967. *A Handbook of Living Primates*. New York: Academic Press.

Napier, P. H., and C. P. Groves. 1983. *Simia fascicularis* Raffles 1821 (Mammalia, Primates): Request for the suppression under the plenary powers of *Simia aygula* Linnaeus, 1758, a senior synonym. Z. N. (S.): 2399. *Bull. Zool. Nom.* 40(2): 117–118.

Nayak, Lahsong N. 1986. Man and the moral order in the Valmiki Ramayana. *Man in India* 66(3): 259–264.

Ohnuki-Tierney, Emiko. 1987. *Monkey as Mirror: Symbolic Transformations in Japanese history.* Princeton, NJ: Princeton University Press.

——. 1990. The monkey as self in Japanese culture. In *Culture Through Time*. Ed. by E. Ohnuki-Tierney. Stanford: Stanford University Press. Pp. 128–153.

Ollier, C. D. 1980. The geologic setting. In *Indonesia: The Making of a Culture*. Ed. by James J. Fox. Canberra, Australia: Australian National University. Pp. 5–19.

Orlove, Benjamin, and S. Brush. 1996. Anthropology and the conservation of biodiversity. *Annual Rev. Anthropol.* 25:329–52.

Palombit, Ryne A. 1992a. A preliminary study of vocal communication in wild long-tailed macaques (*Macaca fascicularis*). I. Vocal repertoire and call emission. *Int. J. Primatol.* 13(2): 143–182.

——. 1992b. A preliminary study of vocal communication in wild long-tailed macaques (*Macaca fascicularis*). II. Potential of calls to regulate intragroup spacing. *Int. J. Primatol.* 13(2): 183–207.

Parker, Sue, and Kathleen Gibson. 1977. Object manipulation, tool use, and sensorimotor intelligence as feeding adaptations in cebus monkeys and great apes. *J. of Human Evolution* 6:623–641.

Peacock, James. 1997. The future of anthropology. *American Anthropologist* 99(1): 9–17.

Peters, E. H. 1986. Grading in the vocal repertoire of Silver Spring rhesus monkeys. In *Primate Ontogeny, Cognition, and Social Behaviour.* Ed. by J. G. Else and P. G. Lee. Cambridge: Cambridge University Press. Pp. 161–168.

Picard, Michel. 1990a. "Cultural tourism" in Bali: Cultural performances as tourist attraction. *Indonesia* 49/50:37–74.

——. 1990b. Creating a new version of paradise. In *Bali, the Emerald Isle*. Ed. by Eric Oey. Lincolnwood, IL: Passport Books. Pp. 68–71.

Pigeaud, Th. G. Th. 1967. Javanese divination and classification. In *Structural Anthropology in the Netherlands*. Ed. by P. E. de Josselin de Jong. The Hague: Martinus Nijhoff. Pp. 61–82.

Pleumarom, Anita. 1994. The political economy of tourism. *Ecologist* 24:142–148.

Pocock, R. I. 1939. The fauna of British India, including Ceylon and Burma. *Mammalia, Vol. 1.* London: Taylor & Francis.

Poirier, Frank, and Euclid Smith. 1974. The crab-eating macaques (*Macaca fascicularis*) of Angaur Island, Palau, Micronesia. *Folia Primatologica* 22:258–306.

Pollmann, Tessel. 1990. Margaret Mead's Balinese: The fitting symbols of the American dream. *Indonesia* 49/50:1–36.

Pollock, Sheldon. 1993. Ramayana and political imagination in India. *J. of Asian Studies* 52(2): 261–297.

Povinelli, D., K. Parks, and M. Novak. 1992. Role reversal by rhesus monkeys but no evidence of empathy. *Animal Behaviour* 44:269–281.

Quiatt, Duane. 1979. Aunts and mothers: Adaptive implications of allomaternal behavior of nonhuman primates. *American Anthropologist* 81:309–319.

Rabor, D. 1968. The present status of the monkey-eating eagle, *Pithecophagia jefferyi Ogilvie-Grant*, of the Philippines. *International Union Conservation Natural Resources Pub. New Series* 10:312–314.

Rassers, W. H. 1925. On the meaning of Javanese drama. In *Panji, The Culture Hero. 1982. Koninklijk Institute Voor Taal-, Land-en Volkenkunde. Translation Series 3.* The Hague: Martinus Nijhoff. Pp. 1–61.

Rees, E. 1960. *The Odyssey of Homer.* Random House.

Richard, A. F., S. J. Goldstein, and R. E. Dewar. 1989. Weed macaques: The evolutionary implications of macaque feeding ecology. *Int. J. Primat.,* 10:569–594.

Richards, Paul W. 1966. *The Tropical Rainforest.* Cambridge: Cambridge University Press.

Rijksen, H.D.A. 1978. A field study on Sumatran orangutans. *Mededelingen Lanbouwhogeschool Wageningen, Nederland* 78–2. Wageningen, Holland: H. Veenman & Zonen B. V.

Rodman, Peter S. 1978. Diets, densities, and distributions of Bornean primates. In *Arboreal Folivores.* Ed. by G. G. Montgomery and F. Eisenberg. Washington, DC: Smithsonian Institution Press. Pp. 465–478.

Rohrich, Lutz. 1991. *Folktales and Reality.* Bloomington: Indiana University Press.

Rubinstein, Raechelle. 1991. The Brahmana according to their babad. In *State and Society in Bali.* Ed. by H. Geertz. Leiden: KITLV Press. Pp. 43–84.

Sankaranarayanan, Kalpakam. 1992. Greater India outside India, with special reference to Indonesia, Bali, and Cambodia. *Journal of the Institute of Asian Studies* 10:3–42.

Santosa, Silvio. 1985. *Gianyar Valley of the Ancient Relics, Art and Culture.* Gianyar, Bali: Regency Government of Gianyar.

Santoso, Soewito. 1980a. *Ramayana Kakawin.* Singapore: Institute of Southeast Asian Studies, and New Delhi: International Academy of Indian Culture, Arya Bharati Mudranalaya.

——. 1980b. The old Javanese Ramayana, its composer and composition. In *The Ramayana Tradition in Asia.* Ed. by V. Raghaven. New Delhi: Sahitya Akademi. Pp. 20–39.

Schreber, Johann Christian. 1774. Die Säugthiere. In *Abbildungen nach der Natur.* Erlangen: Schmetterlingswerkes.

Schultz, Adolph Hans. 1956. Post-embryonic age changes. In *Primatologica Vol. I.* Ed. by A. H. Hofer, A. Schultz, and D. Starck. Basel, Switzerland: S. Karger. Pp. 887–964.

Scott-Kemball, Jeune. 1959. The Kelantan Wayang Siam shadow puppets "Rama" and "Hanuman." *Man* 39(108): 73–78.

Setia, Putu. 1994. Kuta Encounters. In *Travelling to Bali.* Ed. by A. Vickers. Oxford: Oxford University Press. Pp. 71–77.

Seyfarth, Robert M. 1977. A model of social grooming among adult female monkeys. *J. Theor. Biol.* 65:671–698.

Sharon, Douglas. 1993. The metaphysics of curanderismo and its cultural roots. In *Sorcery and Shamanism.* Ed. by D. Joralemon and D. Sharon. Salt Lake City: University of Utah Press. Pp. 165–187.

Shirek-Ellefson, Judith. 1972. Social communication in some old world monkeys and gibbons. In *Primate Patterns.* Ed. by P. Dolhinow. New York: Holt, Rinehart and Winston. Pp. 297–311.

Shively, C., S. Clarke, N. King, S. Schapiro, and G. Mitchell. 1982. Patterns of sexual behavior in male macaques. *Amer. J. Primatol.* 2:373–384.

Silk, Joan B. 1980. Kidnapping and female competition among captive bonnet macaques. *Primates* 21:100–110.

Singh, G. S. 1997. Sacred groves in Western Himalaya: An eco-cultural imperative. *Man in India* 77(2 & 3): 247–257.

Singh, Sheo D. 1969. Urban Monkeys. *Sci. Amer.* 221(1): 108–115.

Smith, Valene L. 1977. Introduction. In *Hosts and Guests: The Anthropology of Tourism*. Ed. by V. L. Smith. Philadelphia: University of Pennsylvania Press. Pp. 1–14.

Sody, H. 1949. Notes on some primates, carnivora, and the babirusa from the Indo-Malayan and Indo-Australian regions. *Treubia* 20:121–90.

Southwick, Charles H., and F. C. Cadigan, Jr. 1972. Population studies of Malaysian primates. *Primates* 13(1): 1–18.

——, and M. Farooq Siddiqi. 1977. Population dynamics of rhesus monkeys in India. In *Primate Conservation*. Ed. by Prince Rainier III and G. Bourne. New York: Academic Press. Pp. 339–362.

——. 1985. The rhesus monkey's fall from grace. *Natural History* 2:62–71.

——. 1994. Primate commensalism: The rhesus monkey in India. *Revue D'Ecologie (Terre & Vie)* 49(3): 223–231.

——, M. Farooq Siddiqi, M. Yahya Farooqui, and Bikas Chandra Pal. 1976. Effects of artificial feeding on aggressive behaviour of rhesus monkeys in India. *Anim. Behav.* 24:11–15.

——, Z. Yongzu, J. Haisheng, L. Zhenhe, and Q. Wenyuan. 1996. Population ecology of rhesus macaques in tropical and temperate habitats in China. In *Evolution and Ecology of Macaque Societies*. Ed. by John E. Fa and Donald G. Lindburg. Cambridge: Cambridge University Press. Pp. 95–105.

Sponsel, Leslie. 1997. The human niche in Amazonia: Explorations in ethnoprimatology. In *New World Primates*. Ed. by W. Kinzey. New York: Aldine De Gruyter. Pp. 143–165.

Sprunger, David. 1996. Parodic animal physicians from the margins of medieval manuscripts. In *Animals in the Middle Ages*. Ed. by Nona C. Flores. New York: Garland. Pp. 67–81.

Stevens, Jane. 1994. Growing rice the old-fashioned way, with computer assist. *Technology Review* 97:16–18.

Struhsaker, Thomas. T. 1969. Correlates of ecology and social organization among African cercopithecines. *Folia primatol.* 11:80–118.

Suchitra, Samanta. 1994. The self-animal. *J. Asian Studies* 53(3): 779–803.

Sukawati, Tjokorda Gde Agung. 1979. Reminiscences of a Balinese prince as dictated to Rosemary Hilbery. *Southeast Asia Paper No. 14*. Honolulu: University of Hawaii, Southeast Asian Studies Publications.

Sussman, Robert W., and Ian Tattersall. 1981. Behavior and ecology of *Macaca fascicularis* in Mauritius: A preliminary study. *Primates* 22(2): 192–205.

———. 1986. Distribution, abundance, and putative ecological strategy of *Macaca fascicularis* on the island of Mauritius, Southwestern Indian Ocean. *Folia primatol.* 46:28–43.

Swellengrebel, J. L. 1984. *Bali: Studies in Life, Thought and Ritual*. Cinnaminson, NJ: Foris Publications.

Swisher, C. C., G. H. Curtis, T. Jacob, A. Getty, A. Suprijo, and Widiasmoro. 1994. Age of the earliest known hominids in Java, Indonesia. *Science* 263:1118–1121.

Taha, Adi Haji. 1985. The re-examination of the rockshelter of Gua Cha, Ulu Kelantan, West Malaysia. *Federation Museums Journal* 30, New Series, p. 134. Kuala Lumpur, Malaysia: Museums Department.

Takahata, Yukio. 1991. Diachronic changes in the dominance relations of adult female Japanese monkeys of the Arashiyama B group. In *The Monkeys of Arashiyama*.

Ed. by Linda M. Fedigan and Pamela J. Asquith. Albany: State University of New York Press. Pp. 123–139.

Takenaka, Osamu. 1986. Blood characteristics of the crab-eating monkeys (*Macaca fascicularis*) in Bali Island, Indonesia: Implications of water deficiencies in West Bali. *Journal of Medical Primatologica* 15:97–104.

Turner, Victor 1967. *The Forest of Symbols*. Ithaca, NY: Cornell University Press.

Uhlenbeck, E. M. 1989. The problem of interpolation on the old Javanese Ramayana Kakawin. *Bijdragen tot de taal-, land-en volkenkunde*. Pp. 324–335.

Valentijn, Francois. 1994 [1724–26]. The slave trade and relations with the Dutch. In *Travelling to Bali*. Ed. by A. Vickers. Singapore: Oxford University Press. Pp. 78–88.

Van Ossenbruggen, F.D.E. 1983. Java's Monca-Pat: Origins of a primitive classification system. In *Structural Anthropology in the Netherlands*. Ed. by P. E. De Joss. Holland: Foris Publications. Pp. 32–60.

Van Schaik, Carel, Maria A. Van Noordwijk, Rob J. de Boer, and Isolde den Tonkelaar. The effect of group size on time budgets and social behaviour in wild long-tailed macaques (*Macaca fascicularis*). *Behav. Ecol. Sociobiol.* 13:173–181.

——, Maria Van Noordwijk, Bambang Warsono, and Edy Sutriono. 1983. Party size and early detection of predators in Sumatran forest primates. *Primates* 2:211–221.

Vickers, Adrian. 1989. *Bali: A Paradise Created*. Berkeley-Singapore: Periplus Editions.

——. 1990. Klungkung regency. In *Bali, the Emerald Isle*. Ed. by Eric Oey. Lincolnwood, IL: Passport Books. Pp. 166–170.

——. 1994. *Travelling to Bali*. Oxford: Oxford University Press.

Vitousek, Peter, Harold Mooney, Jane Lubchenco, and Jerry Melillo. 1997. Human domination of Earth's ecosystems. *Science* 277:494–499.

Walker, D. 1980. The biogeographic setting. In *Indonesia: The Making of a Culture*. Ed. by James J. Fox. Canberra, Australia: Research School of Pacific Studies. Australian National University. Pp. 21–34.

Walker, M. J., and S. Santoso. 1977. Romano-Indian rouletted pottery in Indonesia. *Asian Perspectives* 20(2): 228–235.

Wang, J. 1983. *Taro*. Honolulu: University of Hawaii Press.

Washburn, Sherwood. 1942. The skeletal proportions of langurs and macaques. *Human Biology* 14: 444–472.

——, and Irven DeVore. 1961. Social Life of Baboons. *Sci. Amer.* 204(June): 62–71.

Weiner, Annette. 1995. Culture and our discontents. *Amer. Anthrop.* 97(1): 14–40.

Wessing, Robert. 1978. Cosmology and social behavior in a West Javanese settlement. *Papers in International Studies, Southeast Asia Series No. 47*. Athens: Ohio University Center for International Studies, Southeast Asia Program.

——. 1988. Spirits of the earth and spirits of the water: Chthonic forces in the mountains of west Java. *Asian Folklore Studies* 47:43–61.

Wheatley, Bruce P. 1978a. The behavior and ecology of the crab-eating macaque (*Macaca fascicularis*) in the Kutai Nature Reserve, East Kalimantan, Indonesia. University of California, Davis, unpublished Ph.D. diss.

——. 1978b. Riverine secondary forest in the Kutai Nature Reserve, East Kalimantan, Indonesia. *Malayan Nature Journal* 30(4): 19–29.

——. 1980a. Malaria as a possible selective factor in the speciation of macaques. *Journal of Mammalogy* 61:307–311.

——. 1980b. Feeding and ranging of East Bornean *Macaca fascicularis*. In *The Macaques: Studies in Ecology, Behavior and Evolution*. Ed. by Donald G. Lind-

burg. New York: Van Nostrand Reinhold. Pp. 215–246.
——. 1982. Adult male replacement in *Macaca fascicularis* of East Kalimantan, Indonesia. *International Journal of Primatology* 3(2): 203–219.
——. 1988. Cultural behavior and extractive foraging in *Macaca fascicularis*. *Current Anthropology 29*:516–519.
——. 1989. Diet of Balinese temple monkeys. *Kyoto University Overseas Research Report. Studies on Asian Non-Human Primates* No. 7:62–75.
——, A. Fuentes, and D. K. Harya Putra. 1993. The primates of Bali. *Asian Primates 3*(1 & 2): 1–2.
——, and D. K. Harya Putra. 1994a. The effects of tourism on conservation at the Monkey Forest in Ubud, Bali. *Revue D'Ecologie (Terre & Vie)* 49(3): 245–257.
——. 1994b. Biting the hand that feeds you: Monkeys and tourists in Balinese monkey forests. *Tropical Biodiversity* 2(2): 317–327.
——. 1995. Hanuman, the monkey god leads conservation efforts in the Balinese Monkey Forest at Ubud, Indonesia. *Primate Report* 31:55–64.
——, D. K. Harya Putra, and Mary K. Gonder. 1996. A comparison of wild and food-enhanced long-tailed macaques (*Macaca fascicularis*). In *Evolution and Ecology of Macaque Societies*. Ed. by J. E. Fa and D. G. Lindburg. Cambridge: Cambridge University Press. Pp. 182–206.
White, J. P. 1984. Melanesia. In *Prehistoric Indonesia*. Ed. by Pieter Van De Velde. Cinnaminson, NJ: Foris Publications. Pp. 94–119.
White, Lynn, Jr. 1967. The historical roots of our ecologic crisis. *Science* 155:1203–1207.
Whitmore, T. C. 1975. *Tropical Rainforests of the Far East*. Oxford: Clarendon Press.
Wiener, Margaret J. 1995. *Visible and Invisible Realms*. Chicago: University of Chicago Press.
Wienker, Curtis A., and Kenneth Bennett. 1992. Trends and developments in physical anthropology, 1990–91. *Am. J. Phys. Anthrop*. 87:383–393.
Wikan, Unni. 1989. Managing the heart to brighten face and soul: Emotions in Balinese morality and health care. *American Anthropologist* 16:294–312.
Wiley, R. Haven, and Douglas G. Richards. 1978. Physical constraints on acoustic communication in the atmosphere: Implications for the evolution of animal vocalizations. *Behav. Ecol. Sociobiol.* 3:69–94.
Wilson, Carolyn C., and Wendell L. Wilson. 1975. The influence of selective logging on primates and some other animals in East Kalimantan. *Folia primatol*. 23:245–274.
——. 1977. Behavioral and morphological variation among primate populations in Sumatra. *Yearbook of Phys. Anthrop*. 20:207–233.
Winduwinata, Prijana. 1978. Don't become a teacher. *Indonesia* 25:110–114.
Wittkower, Rudolf. 1942. Marvels of the East. *Journal of the Warburg and Courtauld Institutes* 5:159–197.
Worsley, Peter J. 1972. Babad Buleleng. *Bibliotheca Indonesica, 8*. The Hague: Martinus Nijhoff.
——. 1984. E 74168. *Review of Indonesian and Malayan Studies*. Pp. 65–109.
Wrangham, Richard W. 1980. An ecological model of female-bonded primate groups. *Behaviour* 75:262–300.
——, F. De Waal, and W. C. McGrew. 1994. The challenge of behavioral diversity. In *Chimpanzee Cultures*. Ed. by R. W. Wrangham, W. C. McGrew, F. De Waal, and P. Heltne. Cambridge: Harvard University Press. Pp. 1–18.
Yapp, Brunsdon. 1982. *Birds in Medieval Manuscripts*. New York: Schocken Books.
Yeager, Carey P. 1996. Feeding ecology of the long-tailed macaque (*Macaca fascicularis*)

in *Kalimantan Tengah, Indonesia. Int. J. Primatol.* 17(1): 51–62.
Zhao, Q. K. 1994. A study on semi-commensalism of Tibetan macaques at Mt. Emei, *China. Revue D'Ecologie (Terre & Vie)* 49(3): 259–271.
Zimmer, Heinrich. 1968. *The Art of Indian Asia*. Princeton, NJ: Princeton University Press.
Zoetmulder, P. J. 1974. *Kalangwan*. The Hague: Martinus Nijhoff.
——. 1982. *Old Javanese-English Dictionary*. S-Gravenhage: Martinus Nijhoff.
Zurbuchen, Mary S. 1987. *The Language of Balinese Shadow Theater*. Princeton, NJ: Princeton University Press.

Index